THE COMMONWEALTH AND INTERNATIONAL LIBRARY
Joint Chairmen of the Honorary Editorial Advisory Board
SIR ROBERT ROBINSON, O.M., F.R.S., LONDON
DEAN ATHELSTAN SPILHAUS, MINNESOTA

CHEMICAL INDUSTRY
General Editors: J. DAVIDSON PRATT AND T. F. WEST

Plastics and Synthetic Rubbers

Plastics
and Synthetic Rubbers

BY
A. J. GAIT

AND
E. G. HANCOCK

PERGAMON PRESS

OXFORD · NEW YORK · TORONTO

SYDNEY · BRAUNSCHWEIG

PERGAMON PRESS LTD.,
Headington Hill Hall, Oxford

PERGAMON PRESS INC.,
Maxwell House, Fairview Park, Elmsford, New York 10523

PERGAMON OF CANADA LTD.,
207 Queen's Quay West, Toronto 1

PERGAMON PRESS (AUST.) PTY. LTD.,
19a Boundary Street, Rushcutters Bay, N.S.W. 2011, Australia

VIEWEG & SOHN GMBH,
Burgplatz 1, Braunschweig

First edition 1970

Library of Congress Catalog Card No. 78–116778

Printed in Great Britain by A. Wheaton & Co., Exeter

08 015848 X (flexicover)
08 015849 8 (hard cover)

Contents

Editors' Preface

We were asked by Sir Robert Robinson, o.m., p.p.r.s., to organize the preparation of a series of monographs as teaching manuals for senior students on the Chemical Industry, to be published by Pergamon Press as part of the Commonwealth and International Library of Science, Technology, Engineering and Liberal Studies, of which Sir Robert is Chairman of the Honorary Editorial Advisory Board. Apart from the proviso that they were not intended to be reference books or dictionaries, the authors were free to develop their subject in the manner which appeared to them to be most appropriate.

The first problem was to define the Chemical Industry. Any manufacture in which a chemical change takes place in the material treated might well be classed as "chemical". This definition was obviously too broad as it would include, for example, the production of coal gas and the extraction of metals from their ores; these are not generally regarded as part of the Chemical Industry. We have used a more restricted but still a very wide definition, following broadly the example set in the special report (now out of print) prepared in 1949 by the Association of British Chemical Manufacturers. Within this scope, there will be included monographs on subjects such as coal carbonization products, heavy chemicals, dyestuffs, agricultural chemicals, fine chemicals, medicinal products, explosives, surface active agents, paints and pigments, plastics and man-made fibres.

A list of monographs now available and under preparation is appended.

We wish to acknowledge our indebtedness to Sir Robert Robinson for his wise guidance and to express our sincere appreciation of the encouragement and help which we have received from so many individuals and organizations in the industry.

The lino-cut used for the covers of this series of monographs was designed and cut by Miss N. J. Somerville West, to whom our thanks are due.

J. Davidson Pratt
T. F. West } *Editors*

Preface

THIS book is another in the Chemical Industry series and deals with the Plastics and Synthetic Rubber industries, which have in recent years shown such an outstanding rate of growth.

As in any fast-growing industry the picture is constantly changing, and within the limited scope of the book an attempt has been made to include all the relevant information up to the end of 1968.

Both authors have worked for many years in various fields of the plastics and allied industries and the information has been drawn from their own experience as well as a great variety of published sources. They are particularly indebted to the help and advice given by many friends and colleagues and would especially like to mention the names of Mr. K. H. C. Bessant, Mr. P. G. Croft-White, Mr. C. A. Cutler, Dr. J. J. P. Staudinger, Mr. A. A. K. Whitehouse and Mr V. Wood. Each of these has read chapters on subjects in which he is a specialist and has given the authors the benefit of expert knowledge. To all these and to others who have helped, the authors offer their sincere thanks.

Introduction

WHAT IS A PLASTIC?

Newspaper references to the present time as "the plastics age" and to the modern way of life as "our plastics society" have become common and tend to be accepted with that slightly incredulous tolerance usually accorded journalistic flights of fancy. Relatively few people realize that much of what we call our standard of living and many amenities which we take for granted have been made possible only through the development of modern plastics. Thus, while everyone is familiar with plastics, a short and concise definition is difficult to formulate.

The dictionary defines a plastic as a material which can be moulded or deformed into any desired shape and which may, under the right conditions, retain its new shape indefinitely. Materials of this kind, such as clays, natural waxes and resins, have been known and used for many centuries and it is only just over a century ago that the first synthetic plastic to achieve commercial acceptance, known to most people as celluloid, was produced. Modern plastics broadly conform to the dictionary definition; they are usually solid materials which soften gradually on heating, which may then be moulded to the shape required and will retain that shape on cooling. Within this definition there are wide differences in the properties of the many materials commercially available.

As knowledge of organic chemistry developed during the nineteenth and early twentieth centuries, it was found that some chemical reactions produced resinous materials, more or less similar to the natural resins. At first these materials were discarded as being too intractable for further study and as having no practical value. Gradually, however, it was realized that some of them could be used as substitutes for natural products so that commercial applications for them, as well as analytical methods for investigating their structure, began to be developed.

Much research on the new synthetic resins was carried out in support of the commercial applications but it was soon found that these materials, known scientifically as polymers, were also of great interest to the fundamental scientist. Their study became established as a recognized branch of organic chemistry and was later greatly extended into the physical and inorganic fields.

The words "polymer", "synthetic resin" and "plastics" are in general use by workers in the field covered by this book and it is not possible to differentiate between them. Polymer, or more specifically high polymer, is the more precise term and is used to denote a very large molecule formed by the combination of many thousands of small ones. The term synthetic resin was originally used to denote any synthetic material, not obviously crystalline, with properties similar to natural resins such as shellac. It is now used more frequently as a synonym for the word polymer and, more particularly, for the longer established polymers made by reacting phenol or urea with formaldehyde. Plastics is an omnibus word used generally for the materials from which articles are made; a commercial plastic may consist of virtually pure polymer or it may be mixed with substantial amounts of other materials.

In their applications high polymers (or synthetic resins) fall into two broad groups, the thermosetting resins (thermosets) and the thermoplastics. Materials in the first group undergo a chemical change during the processes of heating and forming to the desired shape so that the final product will no longer soften on heating and cannot be reworked. Thermoplastics, on the other hand, do not undergo a chemical change on heating and can, at least in theory, be resoftened and reworked as often as desired. There is a tendency to use the word resins for the thermo-setting group and plastics for the thermoplastic group but this nomenclature should be avoided. It is preferable to follow the Board of Trade terminology and to use the word resins for all the primary polymers and the word plastics for the compounded (mixed) materials from which articles in the widest sense are made.

There are two applications for high polymers which do not fit readily into the above simple classification but are of great commercial importance; these are synthetic fibres and synthetic rubbers or, to use a more general term, synthetic elastomers. Synthetic fibres are high polymers formed into thin filaments which may then be subjected to the usual processes of the textile and allied industries for production of cloth, ropes, twine and so on. These polymers are usually thermoplastic and most of those in commercial production also have some applications as plastics. The synthetic fibres are described in another volume of this series entitled Man-made fibres and are not dealt with in this book. The term synthetic elastomers is applied to a group of polymers having properties akin to those of natural rubber; their applications depend on these rubber-like properties and some of them may be formed into filaments and used as a special kind of synthetic fibre. The polymers are described in this book but their applications, together with those of natural rubber, really constitute a separate industry and are dealt with only briefly.

HOW ARE PLASTICS MADE?

The production of plastics can be roughly divided into three stages:

1. Production of the raw material or monomer.
2. Polymerization of the monomer to produce a resin.
3. Compounding (mixing) of the resin with fillers, plasticizers, colouring matters and other additives to produce a material suitable for conversion to finished articles.

In some cases there may not be a clear division between one or more of the stages while, in others, each stage may consist of one or more separate and distinct processes.

SCOPE OF THIS BOOK

In the breakdown given above, the first stage is essentially one of chemical manufacture and is not described in detail. Sufficient information on the raw materials and chemical processes used for the production of plastics monomers has been given to define the relationship of the various plastics with one another, and the chart facing this page shows how many important raw materials may be derived from crude petroleum. References for further study are given at the end of this introduction.

The second and third stages listed above comprise the major processes for the conversion of monomers into material suitable for the fabrication of finished articles and are the main subject of this book. The various plastics in commercial production, and some which are still under development, have been classified under the two broad headings already mentioned—thermosets and thermoplastics; within these groups the various plastics have been classified according to chemical type. Modified natural polymers and synthetic elastomers are of great commercial importance but do not fit readily into these two broad divisions; they have, therefore, been dealt with in separate parts of the book. In each chapter the polymerization and compounding processes employed have been described in detail and as much information as possible has been given on the capacity and location of commercial plants.

Finished products are made by a wide variety of processes—largely mechanical—some of which have been specially developed for plastics but many of which are adaptations of well-known fabricating techniques. Whereas many of the polymerization and compounding processes are best carried out on a large scale, calling for heavy capital investment, and are confined to a relatively small number of manufacturers, many of the fabrication processes may be conveniently carried out on a small scale and are ideally suited for exploitation by the small firm. In consequence, a large number of manufacturers is engaged in the conversion of plastic materials into finished articles of all kinds and a full description of this part of the plastics industry in Britain would require a large volume in itself.

The plastics manufacturer is greatly concerned with fabricating methods owing to the close relationship between the physical and chemical properties of his product and the fabricating process in which it is used. Most manufacturers maintain large and expensive customer service departments to assist in solving problems which arise in the application of their products and spend large sums of money on research into fabricating methods. In this book a section has been devoted to a description of the most important fabricating methods in general use and the kind of equipment employed. This is by no means exhaustive and is intended only to highlight the problems of quality and processability which have to be considered in the commercial production of plastics.

Another important area to the plastics manufacturer is that of product testing. Plastics are essentially performance products—that is to say they are judged on the way in which the finished article performs its function. This presents problems since the relationship between physical and chemical properties and performance is often obscure. When plastics are used as replacements for more traditional materials as, for example, in the building industry, there is also the problem of extrapolating short-term performance tests over the life of the product, which may run into decades. In this field also a section has been included to outline the more important test methods and equipment currently in use.

These two sections have been placed early in the book in order to explain the unavoidable references to processing methods in the chapters on individual plastics. They are preceded by a short elementary treatment of polymerization theory.

NOMENCLATURE AND UNITS

All industries rapidly develop their own special vocabulary and the plastics industry is no exception. As this book is intended to give a picture of the industry, many of the terms, especially names of products and intermediates, in common use have been used. Where exact chemical names have been necessary, the rules of nomenclature of the International Union of Pure and Applied Chemistry (I.U.P.A.C.) have been applied. Standard British units of measurement have been used unless stated otherwise; temperatures are in degrees centigrade. A glossary of the special terms and abbreviations used will be found at the end of this chapter.

MANUFACTURERS, OFFICIAL BODIES AND STATISTICS

Names of manufacturers have been given, whenever possible, as each individual plastic is described. Since the number of plastics manufacturers, as distinct from processors, is relatively small, some names recur again and again. Although this book is mainly about the British plastics industry, plastics technology is international and no description of the industry in this country would be complete without reference, also, to the activities of foreign companies competing in world markets. A list of companies mentioned in this book, with brief details about them, is given on pp. 10–13.

There are two official bodies mainly concerned with the plastics industry—the Plastics Institute and the British Plastics Federation. The Plastics Institute caters solely for individuals; there are annual examinations for the Diploma, Graduateship and Associateship of the Institute. Ordinary membership is open to anyone interested in plastics. The British Plastics Federation is the trade association representing the interests of individual companies in negotiations with Government, publicity for the industry and similar public relations matters. Through its special committee on statistics it has done much, in collaboration with the Statistics Department of the Board of Trade, to make available reliable figures on plastics production and consumption.

The rubber industry also has two official bodies which are broadly similar to those of the plastics industry. The Institution of the Rubber Industry is the counterpart of the Plastics Institute while the trade association, the Federation of British Rubber and Allied Manufacturers, has similar functions to the British Plastics Federation.

Individual manufacturers publish only limited information on their own activities, and figures given in this book on plant capacities and on the relative size of producers and users are based partly on published information and partly on indirect estimates. Consequently their accuracy cannot be guaranteed. In any case the industry is undergoing such rapid development that figures of this kind quickly become out of date. The authors believe, however, that they do help to build up an overall picture of the industry which changes more slowly.

HISTORICAL

There are several excellent publications dealing with the historical development of the plastics industry and this introduction concludes with only a brief summary of the events that have led up to the present state of the industry as described in this book.

The plastics industry is usually considered to have begun in 1862 when Parkes in the United Kingdom produced a material, consisting of nitrocellulose compounded with camphor, which he called "Parkesine". Like so many British inventions, its commercial exploitation was left to the United States and the Eastman Kodak company, using the name "Celluloid" for the material, produced a flexible film in 1884 which led to the development of the photographic roll film. Methods of moulding the material were also developed but applications were limited by its ready flammability.

Methods of processing casein to form a material having a superficial resemblance to celluloid were worked out in Germany around 1900, but the product never achieved really large scale production. The next major advance was the commercial development of the phenol-formaldehyde resins. The reactions between phenol and formaldehyde were being studied towards the end of the nineteenth century but it was not until 1910 that Baekeland, in the United States, produced the first commercially useful phenol-formaldehyde resins and explained the principle of making phenol-formaldehyde moulding powder. The product was named "Bakelite" and is still of major industrial importance. The method of modifying these resins to give products compatible with drying oils was worked out by Albert in Germany in 1913–14 and the products became well known as "Albertols".

Urea-formaldehyde resins were developed in both the United Kingdom and Germany about 1920 and melamine-formaldehyde resins rather later. The reactions between glycerol and phthalic anhydride had also been widely studied in the early years of this century but it was not until about 1927 that use of the products in paint manufacture began in Germany.

The resins based on formaldehyde and on phthalic anhydride so far described are thermosets and it was only in the 1930's that extensive commercial development of the major thermoplastics began. That both vinyl chloride and styrene could form some kind of polymer was known in the nineteenth century and polyvinyl chloride (PVC) was characterized as early as 1912. Staudinger published his classic work on the mechanism of polymerization in the late 1920's and early 1930's and this undoubtedly helped to speed up development. Serious production of PVC (in the United States) and of polystyrene (in both the United States and Germany) commenced in the mid-thirties.

In 1934 E. W. Fawcett and his co-workers at I.C.I., using very high pressure techniques for the first time, discovered polyethylene although there was only limited commercial production till after the Second World War. Development of high density polyethylene and polypropylene came later still and was made possible by the work of Ziegler (Germany) and Natta (Italy) on new catalysts in the 1950's.

Work on the polymerization of acrylates began early in this century but development was slow, mainly owing to the high cost of producing the raw materials. I.C.I. devised a direct method of making methyl methacrylate during the late 1920's and commercial production of polymethyl methacrylate started before the Second World War. Academic work on epoxy and polyurethane resins was being carried out in the 1930's but they were not exploited commercially until the 1950's, the former in the United States and the latter in Germany.

Research into the production of synthetic elastomers began before the First World War and there was actually some relatively small scale production of a synthetic rubber based on dimethyl butadiene in Germany in 1917–18. Between the wars, efforts seem to have been directed mainly to development of special purpose rubbers which would avoid the limitations of natural rubber for some applications and substantial progress was made. The production of a commercially acceptable general purpose synthetic rubber was greatly accelerated by the German self-sufficiency programme in the early 1930's and, in the United States, by the entry of Japan into the war and the fall of the Far East.

The search for new plastics is still going on and new uses for the existing ones are constantly being developed, so that the industry is always in a state of change. This book gives an account of the situation, mainly in the United Kingdom, at the time of writing in December 1968. It is intended for senior science students in schools and for university students proceeding to a first degree in a scientific subject. The book assumes that the reader has a knowledge of chemistry to about A-level standard; it will also be useful to non-specialists in industry and commerce who need a general knowledge of the production and uses of plastics and elastomers.

READING LIST

The First Century of Plastics, by M. Kauffman, The Plastics Institute, London, 1963.

Landmarks of the Plastics Industry, Plastics Division, Imperial Chemical Industries Ltd., Welwyn Garden City, Herts, 1962.

Manufacture of Plastics, by W. M. Smith, Reinhold, New York, 1964.

Polymer Technology, by D. C. Miles and J. H. Briston, Temple Press Books, London, 1965.

Polymers and Resins, by D. Golding, Van Nostrand, U.S.A., 1959.

COMPANY INFORMATION

The list given below covers the most important companies mentioned in this book. There are many other companies, both large and small, involved directly and indirectly in plastics manufacture and processing whose activities are less closely connected with the subject matter of the book. There is a good deal of vertical and horizontal integration going on in the industry. Vertical integration means complete or partial fusion between companies concerned with various stages of manufacture between the raw material and the finished product for the final consumer. Similarly, horizontal integration means fusion between companies concerned with similar stages of manufacture.

This integration leads to a large number of cross-links between companies in the industry; for more detailed information on inter-relationships, the student should consult the current edition of *Who Owns Whom* published by O. W. Roskill and Co. Reports Ltd., 14 Great College Street, London S.W.1.

AMERICAN CYANAMID. This is a large Corporation with headquarters in the United States and with subsidiaries all over the world. The U.K. branch is Cyanamid of Great Britain Ltd., and has a factory at Gosport, Hampshire.

BADISCHE ANILIN UND SODAFABRIK. This important West German chemical company was part of the pre-1945 I.G. Farbenindustrie combine and now operates independently. Its headquarters are at Ludwigshafen am Rhein and it plays an important part in many world markets; it is commonly known as Badische or as B.A.S.F.

BAKELITE XYLONITE LTD. A joint company in which the Distillers Company and the Union Carbide Corporation each have a 50% share. It is an example of both vertical and horizontal integration and combines the activities of the former Union Carbide subsidiary, Bakelite Ltd. and most of the activities of B.X. Plastics Ltd., formerly a subsidiary of the Distillers Company.

The company has a number of factories in the United Kingdom and produces phenol-formaldehyde resins, PVC, unsaturated polyesters, cellulose nitrate, cellulose acetate and other plastics allied to these basic materials. As well as selling to other fabricators, the company processes some of these materials to film, sheet, laminates, moulding powders and to finished mouldings for the consumer trade. The abbreviation B.X.L. is often used for the company name.

THE BORDEN CHEMICAL COMPANY (U.K.) LTD. This company was formerly Leicester Lovell Ltd. and is now wholly owned by the Borden Chemical

Corporation of the United States. The company has a factory near Southampton and produces epoxy resins for adhesives and other purposes.

BRITISH CELANESE LTD. ⎱ See Courtaulds.
BRITISH CELLOPHANE LTD. ⎰

BRITISH GEON LTD. Now integrated with B.P. Chemicals (U.K.) Ltd. Originally a 50/50 joint company of the Distillers Co. Ltd. and B. F. Goodrich of the United States. The factory is at Barry in South Wales and produces PVC and acrylonitrile rubber.

BRITISH INDUSTRIAL PLASTICS LTD. This group grew from British Cyanides Ltd., the company which pioneered the development of urea resins in the United Kingdom. It is now a wholly owned subsidiary of Turner and Newall whose main interest is in the production of asbestos and its products. Like B.X.L. it is an example of vertical and horizontal integration and has factories in several parts of the country with a concentration in the Birmingham area.

THE BRITISH OXYGEN COMPANY LTD. The main business of the British Oxygen Co. is the supply of industrial gases but it has, in addition, substantial chemical interests. The name is often abbreviated to B.O.C.

THE BRITISH PETROLEUM COMPANY LTD. This is the well known oil company familiar under its initials, B.P. The company has large chemical interests, managed through a subsidiary company, B.P. Chemicals Ltd. The subsidiary responsible for its U.K. chemical operations is B.P. Chemicals (U.K.) Ltd. and it is this company which took over the major part of the chemical activities of the Distillers Company. A wide range of plastics materials is manufactured including high density polyethylene, polystyrene, PVC, phenol and urea–formaldehyde resins, unsaturated polyesters and acrylonitrile rubber.

THE BUSHING COMPANY LTD. A wholly owned subsidiary of A. Reyrolle & Co. Ltd. whose interests are mainly in the electrical industry. The company manufactures laminated plastics, mainly for electrical purposes, and epoxy resin castings.

COMMERCIAL PLASTICS LTD. A wholly owned subsidiary of Unilever. It manufactures calendered and extruded products, PVC and polyethylene film, high impact polystyrene and polyethylene sheeting.

COURTAULDS LTD. A large group of companies mainly concerned with the textile industry but with some chemical and plastics interests. A wholly owned subsidiary is British Celanese Ltd., a large producer of cellulose acetate. Courtaulds also has a majority holding in British Cellophane which produces film from PVC, polyethylene and polypropylene in addition to its main product, regenerated cellulose film.

THE DISTILLERS COMPANY LTD. This group is best known for its whisky and gin which are famous throughout the world. The group had developed large chemical and plastics interests but disposed of the major part of these to B.P. a few years ago. It still retains a 50% holding in Bakelite Xylonite Ltd.

DOW CHEMICAL COMPANY. This is a large U.S. Corporation, the British subsidiary being Dow Chemical (U.K.) Ltd. There are factories at Barry, in South Wales, producing polystyrene and at King's Lynn, Norfolk, where the products are polystyrene foam and styrene–butadiene latices.

DuPont. This name refers to the world's largest chemical company, E. I. duPont de Nemours and Co. of Wilmington, Delaware, U.S.A. The name has been shortened in various ways from time to time. The spelling adopted here is used throughout this book. The U.K. subsidiary is the DuPont Company (U.K.) Ltd.

JAMES FERGUSON & SONS LTD. A wholly owned subsidiary of Wallpaper Manufacturers Ltd. which is, in turn, a wholly owned subsidiary of the Reed Paper Group. The company produces phenol-formaldehyde resins and moulding powders.

FORMICA LTD. This company is owned jointly by American Cyanamid and the De la Rue Company, perhaps best known as fine printers and manufacturers of playing cards. The company is well known for its decorative laminates.

IMPERIAL CHEMICAL INDUSTRIES LTD. The initials I.C.I. are widely recognized as applying to this large chemical group. The group produces virtually a complete range of plastics materials at one or other of its factories. Two important subsidiaries in the plastics field are Bexford Ltd. (wholly owned) producing cellulose acetate film, especially for photographic purposes, and British Visqueen Ltd. (I.C.I. majority holding) producing polyethylene sheet and film.

THE INTERNATIONAL SYNTHETIC RUBBER COMPANY LTD. This company is owned by a consortium of the major U.K. tyre manufacturers. It has factories at Hythe, near Southampton and at Grangemouth, Scotland. The former produces mainly styrene/butadiene rubber and the latter polybutadiene. Abbreviated to I.S.R.

MICANITE AND INSULATORS COMPANY LTD. A wholly owned subsidiary of the General Electric Company (through Associated Electrical Industries). The company produces laminated plastics, particularly for electrical applications. (*Note.* The General Electric Company of the United Kingdom is not to be confused with General Electric in the United States.)

MONSANTO CHEMICALS LTD. The U.K. subsidiary (partly owned) of the Monsanto Chemical Corporation of the United States. It has plants at Hythe, near Southampton producing polyethylene and at Newport, Monmouthshire, producing polystyrene and ABS.

J. W. ROBERTS LTD. A wholly owned subsidiary of Turner and Newall producing asbestos based laminated plastics.

SCOTT BADER & COMPANY LTD. The works and offices are at Wollaston, near Wellingborough, Northamptonshire. The company manufactures unsaturated polyesters and emulsions of vinyl acetate and its copolymers.

SHELL CHEMICALS (U.K.) LTD. Shell is, of course, the name of one of the largest international oil groups. The group also has world-wide chemical interests, centrally controlled but managed locally by subsidiary companies. The U.K. plastics manufacturing centres are at Carrington, near Manchester, for high and low density polyethylenes, polypropylene and polystyrene and at Stanlow, on the Mersey estuary, for epoxy resins.

STERLING MOULDED MATERIALS LTD. The company's main plants are at Stalybridge, Cheshire, producing phenol-formaldehyde resins, polystyrene and ABS. The company has recently taken over the B.X. plastics polystyrene plant at Manningtree and the Associated Electrical Industries phenol-formaldehyde resins plant at Rugby. Both of these plants are to be moved to Stalybridge in due course.

TUFNOL LTD. A relatively small company producing laminated plastics, primarily for engineering purposes.

UNION CARBIDE CORPORATION. This is a large corporation in the United States, originally concerned mainly with mining and mining machinery. The group has diversified widely into the chemical and plastics industries all over the world. The British subsidiary is Union Carbide Ltd.

GLOSSARY OF
ABBREVIATIONS AND UNUSUAL WORDS

Abietic acid	A complex polycyclic acid, the main constitutent of natural rosin.
ABS	An acrylonitrile/butadiene/styrene terpolymer.
Adduct	A product obtained by mixing two or more compounds which is intermediate between a true chemical compound and a purely physical mixture.
A.K.U.	Dutch firm: Algemene Kunstzijde Unie N.V.
A.S.T.M.	The American Society for Testing Materials.
Blowing	The process of generating small bubbles of gas in a viscous polymer so that it expands to form a foam. To be useful the process must be accompanied by solidification of the polymer.
Blowing agent	A compound which will dissolve in or can be dispersed uniformly throughout a polymer and which can act as a source of gas bubbles.
B.I.P.	British Industrial Plastics Ltd.
B.P.	The British Petroleum Company Ltd.
BR	Polybutadiene rubber.
B.S.I.	The British Standards Institution.
B.X.L.	Bakelite Xylonite Ltd.
CAB	Cellulose acetate-butyrate.
Casting	Usually has the normal meaning when a fluid polymer is poured into a mould and allowed to solidify. Cast film or sheet is made by allowing a thin film of fluid polymer to flow through a slot and solidify, either by air cooling or by contact with a cold surface. The product is distinguished from extruded film or sheet for which the polymer, in a viscous state, is forced through a slot by high pressure.
Chain propagation	The process of growth in chains of molecules in which each molecule, as it adds on to the end of the growing chain, either retains or develops an active free end so that chain growth can continue.

Chain stopper	A compound that will inactivate the ends of a growing chain without becoming active itself so that further chain growth is prevented.
Chain transfer	The reaction between the active end of a growing molecular chain and an inactive molecule in which the chain end becomes inactive and transfers its activity to the other molecule which can then act as the starting point for growth of a new chain.
Cladding	The process of covering one material with a sheet of another by mechanical means, such as rolling together or simultaneous extrusion.
Closed cell	A description applied to a foamed plastic in which the walls of the gas bubbles forming the foam are not broken. Conversely an open cell foam is one in which the walls of the bubbles are broken so that there is intercommunication between the cells.
Compounding	The process of mixing a polymer with colouring matter, curing agents, fillers and similar additives, usually with the aid of heat or solvents and accompanied by limited chemical reaction, to produce a uniform mixture suitable for the next stage of processing by moulding, casting or extrusion.
Curing	A process applied to some polymers in which further chemical reaction and polymerization takes place, usually promoted by addition of chemicals known as curing agents.
D.C.L.	The Distillers Company Limited.
DPP	Diphenylol propane, also known as bisphenol-A.
ECH	Epichlorhydrin.
Encapsulation	A general word with the same meaning as potting (q.v.).
EPT	Ethylene–propylene terpolymer rubber. The third component of the terpolymer may vary.
Extrusion	The forcing of viscous material through a shaped orifice by high pressure followed by solidification of the issuing extrudate.
Fluxing	The controlled heating of a mixture of resin with fillers and other additives in mixing equipment so that the resin softens sufficiently to mix uniformly with the other materials.
ft-lb/in.	Foot-pounds per inch.
Glass transition temperature	The temperature, marked by an abrupt change in the coefficient of cubic expansion, at which a polymer acquires an amorphous structure similar to glass.
Hexa	Hexamethylene tetramine.
High frequency heating	Heating by electric currents induced in a material placed at the centre of a coil of wire through which a high frequency alternating current is flowing.

I.C.I.	Imperial Chemical Industries Ltd.
I.G.	The large German chemical combine (Interessen Gemeinschaft) formed after World War I but broken up by the Allies after the Second World War.
Initiator	A compound which will decompose to form active radicals that can act as the starting points for polymer chain formation. An initiator differs from a catalyst as it cannot be recovered unchanged at the end of the reaction.
IR	Polyisoprene-rubber.
I.S.R.	The International Synthetic Rubber Company Ltd.
I.U.P.A.C.	The International Union for Pure and Applied Chemistry.
lb/ft³	Pounds per cubic foot.
lb/in²	Pounds per square inch.
Linters	The short fibres remaining on cotton seed after the long fibres have been removed by the ginning process.
MDI	4,4-di-isocyanatodiphenylmethane.
m.f.	Melamine formaldehyde.
Mixers	A distinction has to be drawn between mixing rolls, in which material is passed back and forth between heated rolls running at different speeds and container mixers, in which the materials to be mixed are agitated by moving blades or by rotation of the container itself. In practice the most important difference in container mixers lies in the mechanical action that they have on the charge. This may vary from a simple agitation with virtually no mechanical action to a powerful disintegrating action which breaks up the charge as it mixes. There are many hundreds of different designs and a detailed account is not possible but, in general, it may be said that the Werner–Pfleiderer type mix with relatively little shearing action; the Baker–Perkins type are larger and have a more powerful shearing action, while the Banbury type is specially designed to apply a powerful disintegrating action.
Nibs	Small pieces of thermoplastic made by chopping extruded strands into short lengths. Nibs may vary in size according to the properties of the plastic and the use for which they are designed. The aim is to get a product which will not stick in hoppers and conveying equipment and which will soften and flow readily during processing.
p.f.	Phenol-formaldehyde.
Platen	Generally means the work table of a machine tool or the roller in a printing press which presses the paper against the type. In the plastics industry platens are the plates of a hydraulic press between which the mould is compressed.
Potting	The complete enclosure of a component, usually for electrical or electronic apparatus, by a cast block of resin

	which provides protection against corrosion and shock as well as electrical insulation.
Processability	Ugly, but expressive, industrial jargon for ability to pass through standard processes without causing difficulties to produce products of the desired properties.
PVC	Polyvinyl chloride.
SBR	Styrene-butadiene rubber.
Spider	A relatively large piece of metal held in a fixed position in an extruder orifice by three or four thin pieces connecting it to the outer casing, in the same way as a spider is suspended in his web.
Stereoregular	Having the groups attached to each carbon atom of a polymer chain arranged spatially in a regularly repeating pattern.
t/a	Tons per annum.
TDI	Tolylene di-isocyanate.
Tenter frame	A frame used in the textile industry for stretching cloth in the direction of the warp and, by analogy, applied in the plastics industry to frame used for stretching plastic film.
Tons/in²	Tons per square inch.
u.f.	Urea-formaldehyde.
w/v	Weight of solute in 100 parts by volume of solution.
w/w	Weight of solute in 100 parts by weight of solution.

PART I

General

CHAPTER 1

Mechanism of the Production of High Polymers

A GREAT deal has been written on polymerization theory. To attempt a detailed treatment of the subject would be inappropriate in this book. It will be convenient, however, to use expressions such as "polyfunctional molecule" or "free radical mechanism" for the proper understanding of which a knowledge of the basic elements of polymerization theory is necessary. This chapter is intended to outline these basic principles and the reading list at the end of the chapter gives a selection of suitable books for more detailed study.

MONOMERS, POLYMERS AND COPOLYMERS

It has already been noted on p. 2 that a polymer molecule is made up of small molecular units, sometimes called mers. Thus, when ethylene oxide reacts with itself and a trace of water under the right conditions, a polyethylene glycol is formed:

$$n\text{CH}_2\text{---CH}_2 + \text{H}_2\text{O} \longrightarrow \text{HOCH}_2\text{CH}_2(\text{OCH}_2\text{CH}_2)_{n-1}\text{OH}$$
$$\diagdown \diagup$$
$$\text{O}$$

In this case the repeating unit, or mer, is —OCH$_2$CH$_2$— and the compound from which it is formed, in this example ethylene oxide, is the monomer. Repeating units cannot usually exist in the free state. Compounds containing two, three, or four repeating units are known as dimers, trimers and tetramers respectively; polymers contain many repeating units, often tens and even hundreds of thousands.

Homopolymers are polymers in which all the repeating units are identical, although they may be differently arranged, as, for example, in branched chains. Copolymers consist of two different kinds of repeating unit; thus the repeating unit derived from ethylene is $-CH_2CH_2-$ and that from propylene $-CH(CH_3)-$ CH_2-. An ethylene–propylene copolymer will contain some of each, usually randomly arranged. It is possible, however, to produce a copolymer in which the two kinds of repeating unit are arranged in groups and this is known as a block copolymer. Thus, if A and B are the repeating units, the copolymer might be:

$$-A-A-A-A-A-B-B-B-B-B-B-B-B-B$$
$$-A-A-A-A-A-A-A-B-B-B-$$

Terpolymers are polymers containing three different repeating units.

FUNCTIONALITY

The classification of plastics into two major groups—thermosets and thermoplastics—has already been noted. Kienle's concept of functionality is useful in understanding the differences between these two groups and may be briefly explained as follows. When two simple chemicals, such as ethyl alcohol and acetic acid, react together one active centre in each molecule actually takes part in the reaction.

$$CH_3CH_2OH + CH_3COOH \longrightarrow CH_3COOCH_2CH_3 + H_2O$$

The ethyl acetate formed has no more free reactive centres and reaction stops at this stage. Thus, each of the reactants is said to be monofunctional and it can be seen that polymers are not formed. In the same way, when ethyl alcohol reacts with sulphuric acid, two molecules of alcohol react with one of the acid to form diethyl sulphate and this completes the reaction. In this case the sulphuric acid is difunctional but still no polymer is produced. When, however, two difunctional molecules react together, a molecular chain may be started which will have a reactive group at each end and further reaction can take place. For example, terephthalic acid and ethylene glycol each have two reactive groups and may react together to form a long chain polyester.

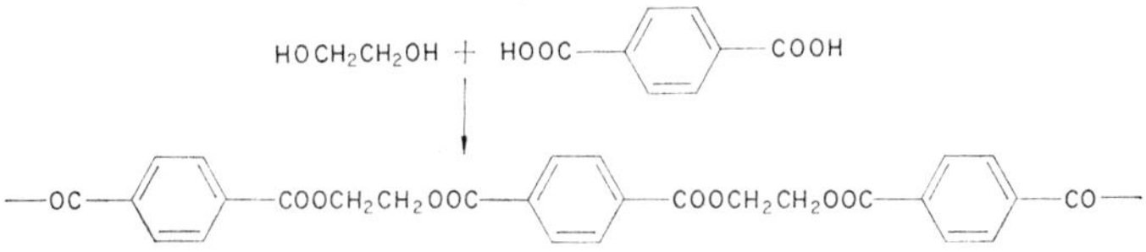

HOCH2CH2OH + HOOC
COOH
OC
COOCH2CH2OOC
COOCH2CH2OOC
CO

This polyester chain will continue to grow until the reaction is terminated by exhaustion of the reactants or saturation of the active groups at the ends of the chain by a monofunctional molecule. Thus, two difunctional compounds reacting together will produce a long-chain polymer. In passing, it may be noted that all mono-olefins, e.g. ethylene and styrene, are difunctional and can, under the right conditions, form chain polymers, although not all of them are commercially useful.

The completed polyester chain described above has no residual reactive centres. If, however, an unsaturated dibasic acid such as maleic acid is used, the olefinic group of the acid is not saturated by the esterification reaction and will give residual reactive points along the polyester chain. The points can be used to link chains together to form a network structure. This cross-linking of chains is usually brought about with the aid of another difunctional compound, such as styrene, and may be represented:

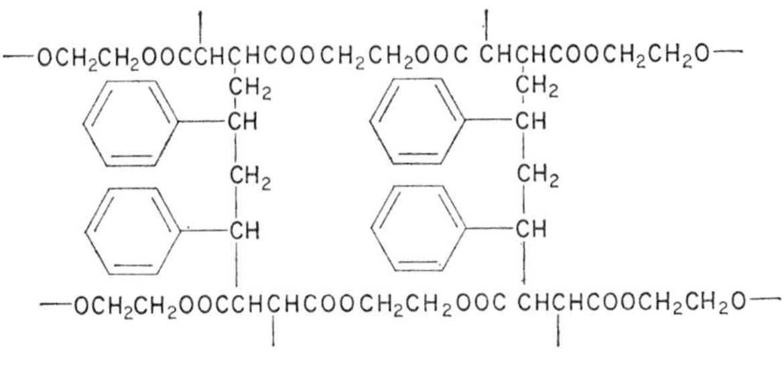

OCH2CH2OOCCHCHCOOCH2CH2OOC CHCHCOOCH2CH2O
CH2
CH
CH2
CH
CH2
CH
CH2
CH
OCH2CH2OOCCHCHCOOCH2CH2OOC CHCHCOOCH2CH2O

In the above example maleic acid has a functionality of four. This network structure is typical of the thermosets and, in the presence of sufficient cross-linking agent, in this case styrene, it will continue to grow until the polymer molecules become so large that they can no longer flow. The resulting material is thus infusible and insoluble in solvents. It can now be seen that, to produce thermosetting resins, at least one of the reacting monomers must have a functionality greater than two.

POLYMERIZATION REACTIONS

In the foregoing section two different kinds of reaction have been described. In the first the reacting molecules simply add together end to end, as in the production of ethylene oxide polymers; this is known as *addition polymerization*. In the second type a small molecule, such as water, is eliminated during the reaction, as in the formation of a polyester; this is known as *condensation polymerization*. It may be noted that thermoplastics are usually produced by addition polymerization while condensation polymers make up the great bulk of commercial thermosets.

ADDITION POLYMERIZATION

The addition of monomer molecules end to end to form a chain polymer can take place by either: (1) a *free radical* mechanism; (2) an *ionic* mechanism.

Free Radical Mechanism

The free radical mechanism is normally limited to monomers containing olefinic linkages, of which ethylene, styrene and vinyl chloride are typical examples. The polymerization is initiated by a compound which can readily undergo decomposition to form a free radical. These compounds are sometimes loosely called catalysts but, since they cannot be recovered unchanged at the end of the reaction, initiator is a more exact name for them. The active free radical first reacts with a monomer molecule, transferring its activity to the monomer portion of the combined molecule and forming the first link in a molecular chain with an active end. Monomer molecules will continue to add on to the activated chain (chain propagation) until growth is terminated by the combination of two chains or by deactivation of the growing chain by an inactive molecule. This may occur through collision of the growing chain with impurities present, including the walls of the containing vessel, or may be brought about by deliberate addition of a compound which acts as a chain stopper.

A good example of this kind of reaction is the polymerization of styrene initiated by benzoyl peroxide. Benzoyl peroxide in solution at temperatures above 60°C will decompose to a short-lived benzoyloxy radical:

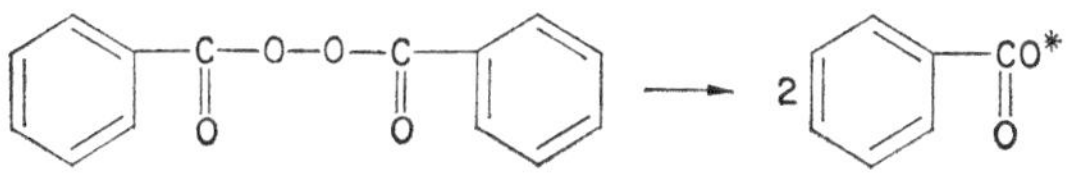

The star represents an unpaired electron at the active point in the molecule. These radicals quickly undergo further decomposition or are neutralized by reaction with the solvent but, in the presence of monomeric styrene, most of them will attach themselves to a styrene molecule and thus initiate a large number of growing chains.

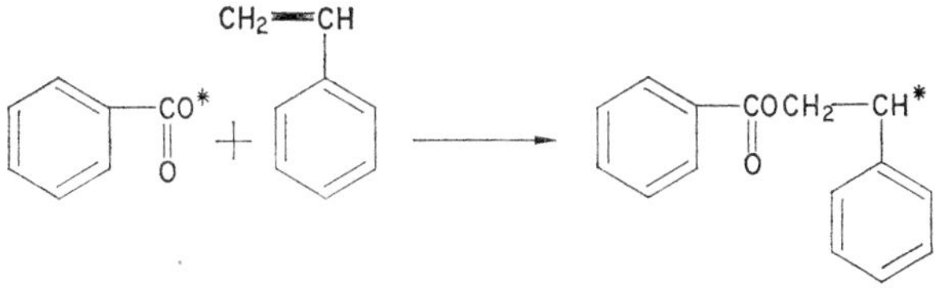

Here the active end is transferred to the styrene and it is ready to add another styrene molecule to form

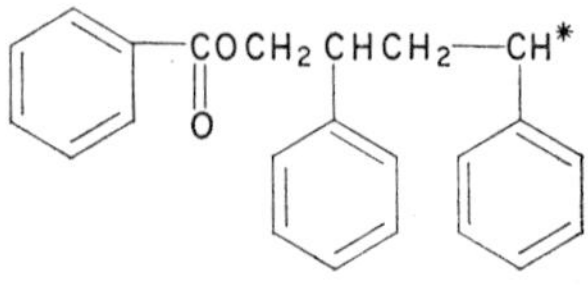

Chain propagation will continue in this way until a molecular weight of many tens, or even hundreds, of thousands has been reached. The reaction is terminated by two chains reacting together, so that the active ends are neutralized, or by the active chain transferring its activity to an inactive molecule. The first reaction has the effect of producing two polymer molecules—one saturated and the other with an olefinic linkage at the end of the chain. For polystyrene this may be represented:

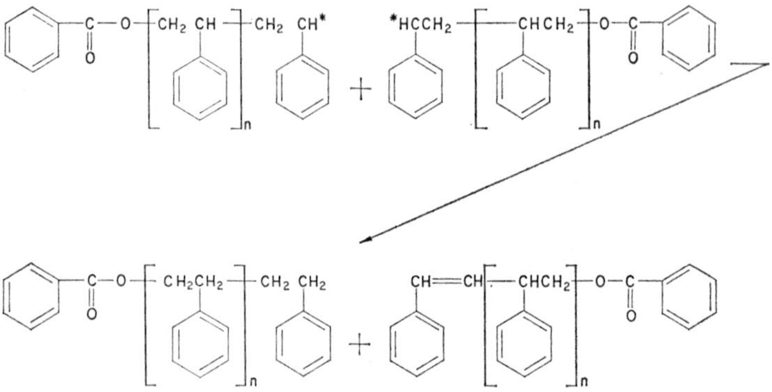

In the second reaction a hydrogen atom is transferred from the inactive molecule to saturate the end of the polymer chain, leaving the originally inactive molecule with an active end capable of starting another polymer chain. This process is often called *chain transfer*.

The peroxide type of initiator described above has to be heated to at least 60°C in order to produce an adequate number of free radicals to start the reaction. In some cases it is necessary to carry out polymerizations at much lower temperatures, when a redox system provides an efficient initiator. A redox system normally contains an inorganic oxidizing agent and a reducing agent; examples of oxidizing agents are ammonium persulphate (also used as a free radical initiator by itself), hydrogen peroxide and potassium chlorate, while typical reducing agents are sodium bisulphite, ferrous sulphate and sodium thiosulphate. As an actual example, the active radical using hydrogen peroxide and ferrous sulphate is the HO radical formed as follows:

$$H_2O_2 + Fe^{++} \longrightarrow OH^- + HO^* + Fe^{+++}$$

This redox system only works in a slightly acid medium and is particularly suitable for monomers which are at least partly water soluble. It can be applied to virtually water insoluble monomers, however, when the polymerization is carried out in an emulsion (see p. 183). It is important to note that, in practice, the two components of a redox system are added to the monomer separately as the active radicals produced in the reaction have only a very short life.

There are several other ways of initiating a free radical polymerization; for example active molecules can be produced in some monomers simply by heating or by irradiation with ultra-violet light. Once the activated chain has been started, it is propagated and terminated by the mechanisms described above.

Ionic Mechanism

Some olefins, such as propylene and isobutylene, do not form high molecular weight polymers with free radical initiators, due to the predominance of chain transfer and side reactions. High polymers can however, be produced using ionic initiators which can also be used to polymerize monomers with a non-olefinic double bond, such as carbonyl compounds. Ionic initiators are said to be anionic or cationic according to whether the end of the growing polymer chain is negatively or positively charged. Typical anionic initiators are alkali metals, n-butyl lithium and sodium methoxide; conventional strong acids are good cationic initiators as are the Friedel–Crafts alkylation catalysts, such as aluminium trichloride and boron trifluoride, when promoted by small quantities of other compounds such as hydrochloric acid or water. As might be expected in an ionic system, the nature of the solvent may exert a profound influence on the course of the reaction; the precise mode of action of these initiators is often not clear. The action of boron trifluoride on isobutene in the presence of traces of water may be represented as follows:

$$BF_3 + H_2O \longrightarrow BF_3OH^- + H^+$$

$$\underset{\text{(the monomer)}}{CH_2=CHR} + H^+ + BF_3OH^- \longrightarrow \underset{\text{(an \textit{ionic} } BF_3 \text{ complex)}}{CH_3—CHR^+BF_3OH^-}$$

$$CH_3CHR^+BF_3OH^- + CH_2=CHR \longrightarrow$$

$$CH_3—\overset{\displaystyle R}{\underset{\displaystyle H}{\overset{|}{\underset{|}{C}}}}—CH_2—\overset{\displaystyle R}{\underset{\displaystyle H}{\overset{|}{\underset{|}{C}}}}^+BF_3\ OH^-$$

With pure materials chain growth can continue in this way until very high molecular weights are attained. Since all chain ends have the same charge, mutual combination of chains is impossible and chains are only terminated by transfer of a hydrogen ion to an impurity or to another monomer molecule.

ZIEGLER TYPE CATALYSTS

There is an important group of catalysts, stemming from work by Professor Karl Ziegler in Germany and extended by Professor Natta in Italy, which does not fit readily into the classifications given above. Ziegler discovered that a mixture of aluminium triethyl and titanium tetrachloride in an inert hydrocarbon solution forms a black precipitate which can initiate the polymerization of ethylene at low pressure. The reaction was quickly extended to olefins in general and to dienes; many other similar catalyst systems have since been developed. An important feature of these catalysts is that they control the spatial arrangement of the monomer molecules during polymerization; this has made possible the commercial synthesis of stereo-regular polymers and, in particular, the synthetic duplication of some natural polymers, such as rubber, which had not previously been possible. The mode of action of these catalysts has still not been fully explained and will not be further dealt with here. Some examples of their commercial applications are given in Chapters 8 and 13.

CONDENSATION POLYMERIZATION

The addition polymerizations described in the previous section proceed extremely rapidly, whereas condensation polymerization, which involves the elimination of a small molecule, usually water, at each stage is relatively slow. In condensation reactions with difunctional reactants the first stage of reaction forms a product with mutually reactive end groups. Thus, in the polyesterification reaction described on p. 21 the reaction between one molecule each of a diol and a diacid will produce a monoester with an OH group at one end and a COOH group at the other. If the spatial arrangement of the molecule is such that these two groups approach one another, then ring closure may occur and polymerization will not proceed. Chains of six carbon atoms are most likely to form rings in this way; rings with more or less atoms are possible but are not formed so readily. If both condensing molecules are large, the first condensation product will have a chain long enough to ensure that its ends are far apart so that chain formation is the favoured reaction and polymers are formed. Thus, in the production of nylon from hexamethylenediamine and adipic acid, reaction between one molecule of each produces a chain of twelve carbon atoms where ring closure is improbable and long chain polymers are formed.

Condensation polymerizations are often carried out under relatively stringent reaction conditions which may cause an active monomer to attach itself at any moderately active point along the chain rather than add on to the end. Chain formation can then continue from the free end of this monomer and a branched chain polymer will be formed.

MOLECULAR WEIGHT OF POLYMERS

It will be clear from the foregoing treatment of polymerization reactions that polymer molecules will vary in size because some will be terminated sooner than others; the molecular weight of a polymer is, therefore, an average. There are two important measures of average molecular weight, viz. the number average and the weight average. If all molecules were the same size the two quantities would be identical; the difference between them increases as the spread of molecular weight between the largest and the smallest molecules increases. Number average is the simplest to understand—it is the arithmetic mean molecular weight and is obtained by dividing the sum of the weights of each molecule by the number of molecules. The weight average molecular weight may be defined as the sum of the fractional weights that each molecule contributes to the average according to the ratio of its weight to that of the whole sample. It is obtained by dividing the sum of the squares of the weights of each molecule by the weight of the whole sample.

Thus, in a sample consisting of six polymer molecules of weights 3, 4, 5, 6, 7, 8 the number average would be:

$$\frac{3 + 4 + 5 + 6 + 7 + 8}{6} = \frac{33}{6} = 5 \cdot 5,$$

and the weight average would be:

$$\frac{9 + 16 + 25 + 36 + 49 + 64}{3 + 4 + 5 + 6 + 7 + 8} = \frac{199}{33} = 6 \cdot 0.$$

The ratio weight average–number average is, therefore a useful measure of the molecular weight spread; it is one when all the molecules are the same size and increases with the spread.

The properties of a polymer are greatly influenced both by its average molecular weight and by the molecular weight spread. For a given polymer, increase in molecular weight increases the melting point, impact strength, melt viscosity and resistance to chemical attack and decreases the solubility of the polymer in solvents; widening the spread of molecular weight generally leads to increased elasticity of the polymer and narrowing it to increased toughness. The determination of physical properties which give a measure of these quantities is, therefore, an important part of plastics testing; the methods used are more fully described in Chapter 2, but it may be noted at this point that, of the common methods of determining molecular weight, number average is given by elevation of boiling point, depression of freezing point, osmotic pressure and end-group analysis which, in effect, count the molecules, and weight average by light scattering and sedimentation velocity measurements which estimate their size.

It must again be emphasized that this chapter gives only a greatly simplified treatment of polymerization theory; some further references to the problems of polymer structure, particularly those of stereoisomerism, are made in the later chapters but the student is referred to the reading list below for detailed study.

READING LIST

An Introduction to Polymer Chemistry, by W. R. Moore, University of London Press, London, 1963.
Organic Chemistry of High Polymers, by R. W. Lenz, Interscience, New York, 1967.
Copolymerization, edited by George E. Ham, Interscience, New York, 1964.
Polymerization and Polycondensation Processes, No. 34, Advances in Chemistry Series, American Chemical Society, Washington, D.C., 1962.

CHAPTER 2

Processing and Testing Methods

In the chapters which follow frequent reference is made to the processing methods and properties of plastics materials. Specialized methods, such as production of foamed materials and of glass reinforced polyesters, are described in the appropriate chapter, but many methods, both of processing and testing, are in wide use throughout the plastics and rubber industries and can conveniently be dealt with in a preliminary chapter.

SECTION A. PROCESSING

The primary processing of plastics involves first heating the material until it is liquid or semi-solid, forming it into the required shape and then solidifying it again, by chemical curing in the case of thermosets and by simple cooling in the case of thermoplastics. The primary shaping processes may be followed by secondary conversion processes, such as machining of formed shapes and making up of films or coated textiles, which differ little from those used in industry generally.

Although relatively simple to describe, the processing of plastics is a highly complex matter in which many factors, chemical, physical and mechanical, play an important part. The maintenance of successful operations and the diagnosis of faults that arise calls for experience and skill. This chapter can do no more than explain briefly the principles involved; suggestions for further reading are given on p. 46.

Compression Moulding

This is one of the oldest methods of converting polymers to useful articles; it is applied mainly to thermosets and to rubbers, which behave like thermosets in some respects. In its simplest form the operation consists of filling one half of a two-piece mould with the chosen plastic in the form of a moulding powder and bringing the two halves of the mould together. The assembly is then fitted into a hydraulic press, with platens which can be heated by steam or electricity to a temperature of 150–180°C, and subjected to a pressure between 500 lb/in² and 3 tons/in².

Moulds are made of high quality steel, chromium plated, or of stainless steel with highly polished internal surfaces to give a good finish to the moulding. Presses are rated according to the total force which they can bring to bear; thus a mould covering an area of 50 in², used in a 100 ton press, will be subject to a pressure of 2 tons/in². The sequence of operations is shown diagrammatically in Fig. 1.

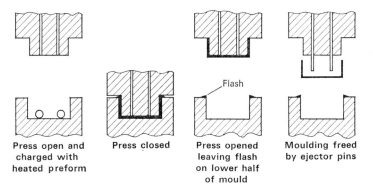

FIG. 1. Simple compression moulding.

The mould shown is known as a flash mould and is filled with a slight excess of moulding powder; when the two halves of the mould are brought together under heat and pressure, the moulding powder softens and flows to take the shape of the mould while the flash or spew is squeezed out at the edges. In some operations the mould is opened momentarily after the first application of pressure to allow evolved gases to escape. The presence of flash ensures that the mould is completely filled, since under-filling would result in weak mouldings of low density; flash represents a waste of material and is, therefore, kept as small as possible. For some deep mouldings it is necessary to use a positive mould in which the plunger penetrates right into the mould so that there is no room for flash. This calls for a moulding powder of specially uniform quality and for very accurate control of the amount of powder charged to the mould.

When the moulding powder begins to soften and flow, it also begins to cure; flow properties and curing time must be related so that the mould is completely filled before the cure has proceeded too far, while the time awaiting completion of the cure, once the mould has been filled, should be at a minimum. In practice curing times will be 1 min or less for small mouldings but as high as 30 min for large mouldings.

Although moulds may be filled, pressed and emptied by hand, moulding is essentially a repetitive operation calling for the production of large numbers of identical articles. In commercial operations, therefore, moulds are usually set up in presses so that they may be filled, closed, opened and the moulded article ejected automatically. The word "tool" is often used in the plastics industry to denote that part of the processing equipment which determines the size and shape of the article being produced; the operation of fitting moulds into presses and preparing them for production runs is known as "tooling up". The moulding cycle may also be speeded up by pre-forming the charge of powder to a suitable shape, usually a ring, and bringing it to near moulding temperature in a high frequency heater, which heats the preform uniformly throughout its bulk; alternatively, preforms may be heated in an oven.

Simple compression moulding as described above has a number of practical limitations. For small articles multi-cavity moulds must be used to obtain an economic production rate and it is difficult to ensure that each cavity receives an equal charge of powder; also many mouldings, especially of electrical components such as sockets, plugs and lamp-holders, require metal inserts round which the softened powder must flow properly before it cures. Many of these problems are overcome by a modified technique known as transfer moulding.

Transfer Moulding

This process is somewhat similar to the injection moulding technique used for thermoplastics. The charge of moulding powder is first softened in a heated chamber and is then forced by the ram of the press through a narrow passage, called a gate, into

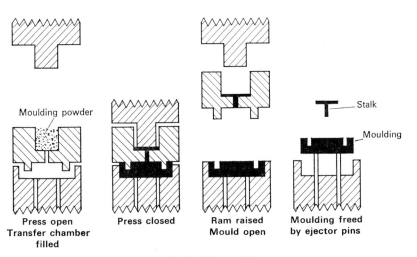

FIG. 2. Transfer moulding

the mould itself. The various stages are illustrated in Fig. 2 and a diagrammatic section through a commercial press in Fig. 3.

The method is particularly suitable for intricate mouldings and for the production of a number of small mouldings with one stroke of the press. Since the plastic is already fluid when it enters the mould, it will penetrate small cavities and there will be less development of localized high-pressure spots as it flows round core pins and inserts. The moulds, too, will be subject to less wear as they do not have to stand the shearing forces developed during opening and closing. Generally, a moulding powder with better flow properties is needed than for simple compression moulding and these properties must be built into the moulding powder formulation.

A disadvantage of the method is that the material left in the flow passages, the "stalk", also cures and is wasted. Production of satisfactory mouldings depends, in large measure, on proper mould design and especially on the correct design and placing of gates; this is a matter calling for specialized experience and skill.

Injection Moulding

This method is especially well adapted to the production of articles from thermoplastic materials, and mouldings varying in weight from an ounce or so to several pounds may be produced at a high rate on automatically controlled moulding machines. Considerable mechanical equipment is required and, because moulding pressures are quite high, the larger machines must be of massive construction and are expensive. Nevertheless, the unit cost may still be quite low provided a run of some thousands of units can be guaranteed.

The method is very simple in principle; it merely involves heating the moulding powder or pellets until the material is sufficiently fluid to be forced through a jet into a cold mould or tool where it sets to the required shape. The process is a cyclical one and a complete cycle, which produces one moulding, or set of mouldings if a multi-cavity mould is being used, may be regarded as a number of separate operations, some of which may be going on simultaneously.

A charge of moulding powder, fed from a hopper, is moved forward into a heating zone in the moulding machine; this may be done continuously by means of a screw or intermittently by

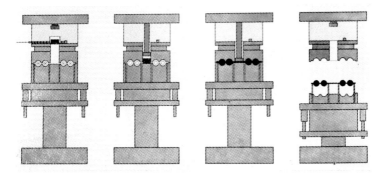

FIG. 3. Transfer moulding—diagram of a typical commercial press.

means of a plunger which pushes a predetermined charge forward at each stroke. In either case the material is gradually softened and then passes into a pre-injection chamber; from this it is driven by one stroke of a ram and passes through a jet into the closed and locked mould. The ram and jet are withdrawn, the mould is opened, the article ejected, the mould closed and locked and the whole cycle is repeated. The operation is illustrated diagrammatically in Fig. 4.

An essential part of the system is that room must be provided for the mould to be opened and the article ejected. This may be achieved by moving one half of the mould back and leaving the other half connected to the injection system; an alternative method is to arrange the injection system so that it can be moved forward and back as the mould closes and opens. Both methods are used in practice; in either case the mechanical equipment is so arranged that the action of forcing a charge into the mould and withdrawing the ram brings a further charge of material forward ready for the next stroke.

Thermosets are occasionally injection moulded and, for these, the extent and duration of pre-heating must obviously be limited. The mechanical effort of moving the material forward through the softening zone may provide sufficient heat to soften it; the mould must, however, be heated to provide for curing the material and the moulded article is ejected hot. Substantial quantities of rubber products are produced by injection moulding and, in this case, heat is provided in both the softening section and in the mould.

In all injection moulding operations temperature control is of the utmost importance. All thermoplastics are liable to heat degradation if held at elevated temperatures for too long and, for materials like PVC, it is essential that the time for which it is held at moulding temperature should be at a minimum. Because of the ease and accuracy with which it can be controlled, electrical heating is generally used. Moulds are cooled by circulating a cooling fluid through built-in channels. A versatile injection moulding machine must be capable of having the temperature

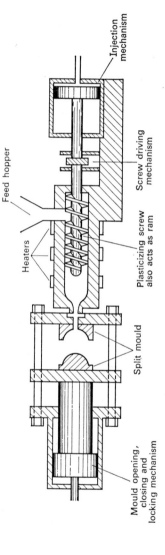

FIG. 4. Sectional diagram of injection moulding machine.

conditions in the system varied to suit the different materials which may be moulded on it. Modern machines are automatically controlled and, once set up for a run, require no attention other than replenishment of the feed hopper and regular inspection of the output to ensure that no faults have developed.

Extrusion

Extrusion is a continuous process in which softened plastic material is forced through a shaped die from which it may emerge in almost any form, from a simple circular rod or tube to a wide,

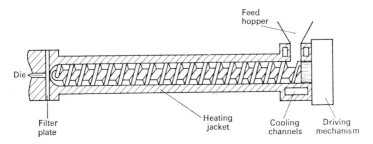

FIG. 5. Single screw extruder.

flat sheet. The softening zone of an extruder is similar to that of an injection moulding machine, with the temperature being gradually raised as the material moves along the barrel. As continuous pressure is required, the driving force is provided by a screw which may also provide for mastication and compounding of the polymer and for part of the heat required through the mechanical work done. The process is shown diagrammatically in Fig. 5.

As for injection moulding, the temperature at the various stages must be carefully controlled. The screw itself may have a hollow core so that it can be heated or cooled and its pitch must be designed to suit the viscosity and other characteristics of the plastics material being used. It is of the utmost importance that the flow through the system should be streamlined so that there are no dead pockets in which softened material can collect and decompose.

Extruders are usually designed so that the die may be changed to suit the shape required. A circular rod obviously needs a die with a circular hole while a tube needs a similar die with a mandrel suspended in the centre of the hole by means of a spider. The die must be long enough to allow the plastic to flow round the legs of the spider and reform a tube. The arrangement is shown in Fig. 6.

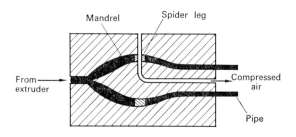

Fig. 6. Diagram of die and mandrel for tube extrusion.

By using a hollow mandrel supplied with compressed air through one of the spider legs, pressure may be applied inside the extruded pipe provided its end is plugged. The extruded pipe is passed through a cooled calibrating tube as it leaves the die and is circumferentially stretched so that its outside diameter is accurately sized to the inside diameter of the calibrating tube. The wall thickness is controlled by the size of the annular space between the mandrel and the die and by the degree of circumferential stretching.

When extruding profile shapes there is considerable surface drag as the plastic passes through the narrower parts of the die, so that the shape of the orifice may be quite different from the profile shape required. An ordinary square bar, for example, requires an orifice with curved sides as shown in Fig. 7.

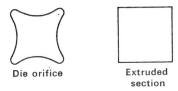

Die orifice

Extruded section

FIG. 7. Relation between shape of die orifice and of extruded section.

The take-off arrangements are an important part of extrusion. Cooling must be applied as soon as the material leaves the die, so that it does not lose its shape; this is usually done by passing the extrudate through a cooling bath but blowing with cold air is sometimes sufficient. The extruded shape is carried on some form of mechanical conveyor until it is either cut into suitable lengths or, if sufficiently flexible, wound on to a reel.

Sheathing wire with plastic is a special kind of tube extrusion. In this case the wire passes continuously through the centre of the die and acts as a mandrel, while the plastic is extruded round it. A typical arrangement for such an extrusion head is shown in Fig. 8.

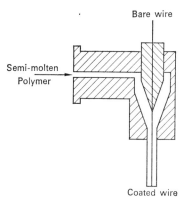

Bare wire

Semi-molten Polymer

Coated wire

FIG. 8. Extrusion coating of wire.

Another modification of tube extrusion is described on p. 171 for the manufacture of lay-flat tubing. Although mainly applied to polyethylene this technique may also be used for unplasticized PVC.

Extruded Film and Sheet

By using a die in the form of a slot a flat film or sheet may be extruded; material $0 \cdot 01$ in. thick or less is known as film while thicker material is designated sheet. Because the width of a slot die is large compared with the diameter of an extruder barrel, the flow paths to the edges are much longer than those to the centre of the die and mechanical arrangements must be made to compensate for this.

Cooling must be applied immediately after extrusion, either by extruding into a cold water bath or into the nip of a pair of highly polished, water-cooled rolls. The latter method is now generally preferred and will produce film and sheet of excellent clarity and gloss.

Thermal Forming

This process is applied particularly to plastics in sheet form. A piece of sheet of the required size is placed over a box containing the mould and is heated, usually by infrared radiation, until soft; vacuum is then applied through the mould, causing the sheet to be sucked tightly against it and to take its shape. The article may be ejected by reversing the air stream and blowing it off. The method is shown in principle in Fig. 9.

There are numerous mechanical variants to the process; plugs may be used in various ways to assist in the major deformation of the sheet which may then be forced into close contact with the mould by vacuum, by air pressure or by a combination of the two. The technique is particularly suitable for the production of mouldings with a large area; the moulds are cheap and may even be of wood or plaster for short runs or for production of

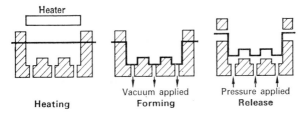

Heater

Heating | Vacuum applied Forming | Pressure applied Release

Fig. 9. Vacuum thermoforming.

prototypes. A commonly seen application of the method is the fixing of small components to showcards by means of a formed transparent sheet. Most thermoplastics may be thermoformed but the ones most commonly used are toughened polystyrene, cellulose acetate and polymethylmethacrylate. The last-named material is relatively expensive but its high clarity and gloss make it specially suitable for advertising signs and similar applications.

Blow Moulding

This technique is somewhat analogous to that of glass-blowing and is applied particularly to production of hollow articles; two methods are in common use. In injection blow moulding a thick-walled tube is first injection-moulded round a blowing stick which is then transferred to a blowing mould; compressed air is passed down the blowing stick and expands the tube to the shape of the mould. In the extrusion method a length of plastic tube is extruded into a split mould which is then closed round a blowing tube, effectively sealing both ends of the trapped piece of tube: this is then expanded to the shape of the mould by means of compressed air. In both cases the moulded article is cooled by contact with the cold mould. The process may be automatically controlled, permitting long runs of standard articles to be produced with a minimum of manual interference. The principles of both methods are illustrated in Figs. 10 and 11.

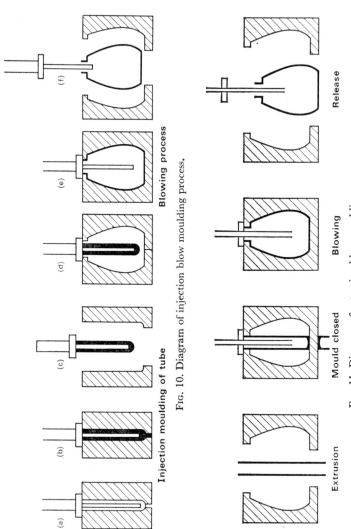

Injection moulding of tube

Blowing process

FIG. 10. Diagram of injection blow moulding process.

Extrusion   Mould closed   Blowing   Release

FIG. 11. Diagram of extrusion blow moulding process.

Rotary Casting

This process, also known as slush moulding, is another method of forming hollow articles; PVC in the form of a plastisol (see p. 213) is the most commonly used plastic.

The hollow moulds, usually of metal, are open at the top and are first heated to about 108°C. A quantity of plastisol is poured in and the mould given a swirling motion until the plastisol begins to gel; after a minute or so any excess plastisol is poured out and the mould held at about 150°C for about 5 min to complete the gelling process. The flexible article may be removed by collapsing it under a vacuum or the mould may be designed so that it can be split.

Calendering

This is another method of producing plastic sheet; it is a very important process, originally developed in the rubber industry and adapted for plastics. A calender consists essentially of one or more pairs of rotating hollow rolls which may be heated internally by steam or hot water. The compounded plastic is fed into the nip of the first pair of rolls and emerges from the last pair in sheet form; the surface may be plain or embossed and is determined by the surface of the last roll.

Calenders are massive and expensive machines and many mechanical problems arise in ensuring that they produce sheet of uniform thickness and quality; these are outside the scope of this chapter.

Coating

By means of coating techniques it is frequently possible to confer the desirable properties of plastics on the coated material without impairing the characteristics of the substrate. Textiles, paper and metals may be coated in a variety of ways.

Calendering Technique

This is largely used for coating textiles with PVC and, to a lesser extent, with polyethylene. A calendered sheet of the plastic is brought alongside a textile cloth, normally a thin material of the same width, and the two sheets pass between hot rolls. The upper roll in contact with the plastic moves somewhat faster than the lower one in contact with the cloth; this forces the plastic into the interstices of the cloth and a good bond is formed. The coated cloth is cooled and rolled up. Very large quantities of leathercloth are made in this way by coating fabrics with PVC.

Paper may be coated in a somewhat similar way. An extruded plastic sheet is brought into contact with a sheet of paper while still hot and the two are passed between a large, chilled roll and a smaller pressure, idling roll. The coated paper leaves the chilled roll in a form in which it can be wound up ready for use. The process is shown diagrammatically in Fig. 12. Plate I shows a plastic sheet being extruded on to a paper substrate as it passes into the nip of a pair of rolls.

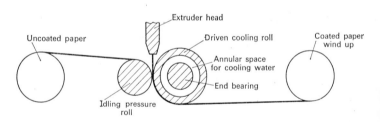

Fig. 12. Coating paper with a thermoplastic.

Spreading Techniques

These are used for applying a thick coating to a reasonably heavy textile base; this may have several layers and be $\frac{1}{4}$ in. or so thick. In a typical method a PVC plastisol is applied across the surface of the moving textile base which then passes under a spreader knife which ensures that the plastisol is spread evenly

PLATE I. Coating a paper substrate with a plastic film. (By courtesy of British Resin Products Ltd.)

and leaves a smooth surface. The coated textile then passes through a long infrared oven in which the heating is controlled to bring the temperature up to the gelling point of the plastisol; when gelation is complete the material is cooled. Cloth up to 48 in. wide can be coated by this method, which is widely used in the manufacture of conveyor belts.

Coating of Metals

Metals may be coated with a film of plastic, both to improve their appearance and to protect them from corrosion. Large surfaces such as the insides of tanks may be lined with plastic sheets which can be joined together by welding processes. Thick sheets are joined at the edges in the same way as metal sheets by using a stream of hot gas as a welding torch. Thinner sheets are overlapped and pressed together under heat supplied by a hot tool or a stream of hot gas. For small metal containers such as drums a loose lining to fit the inside of the drum is sometimes used.

Plastics may be sprayed to form a coherent coating by blowing finely divided polymer particles through a heated nozzle, using a compressed gas as the driving force. This method is useful for coating large surfaces without complicated nooks and crannies.

Small metal articles are usually dip coated. For PVC the preheated article is dipped into a PVC plastisol, when a layer of plastisol gels on the hot metal; excess plastisol is allowed to run off and the article given a further heat treatment in an oven to complete the gelation of the PVC layer. For other plastics a bed of finely divided polymer is fluidized by having a stream of gas passed continuously through it and the preheated articles are dipped into this fluidized bed. A layer of particles adheres to the hot metal and is fused into a coherent coating by subsequent heating in an oven.

In all of these coating processes involving heating care has to be taken to control temperatures and avoid degradation of the plastic.

Plastics Processing In Industry

Plastics processing is spread widely throughout industry. It may be carried out in small establishments having one or two moulding presses, in large factories producing a wide range of plastic goods or as an adjunct to the main manufacturing process for all kinds of products. Many modifications to the general processes described have been devised to suit particular applications.

The manufacture of plastics processing machinery is quite a substantial industry in itself and of fundamental importance to the solution of many processing problems.

READING LIST

Compression and Transfer Moulding, by A. Butler, Iliffe Press' or the Plastics Institute, London, 1964.
Plastics Mould Design, Vol. 1, *Compression and Transfer Moulds*, by R. H. Bebb, Iliffe Press for the Plastics Institute, London, 1964.
Extrusion of Plastics, by E. G. Fisher, Iliffe Press for the Plastics Institute, London, 1964.
Injection Moulding of Plastics, by J. S. Walker and E. R. Martin, Iliffe Press for the Plastics Institute, London, 1966.
Blow Moulding, by D. A. Jones and T. W. Mullen, Reinhold, New York, 1961.
Plastics Forming, by R. L. Butzko, Reinhold, New York, 1958.

SECTION B. TESTING

The ultimate test for a plastic is its performance in the chosen application. A great deal of performance testing to destruction is carried out, especially when developing new applications. For effective control it is necessary to relate these tests to the results of non-destructive testing of manufactured products and, more important still, to physical or chemical properties of the compounded plastic and of the resin itself. The subject is a large one on which books have been written, some of which are listed on p. 56; in this section it is only possible to give brief accounts of the methods used to determine the molecular weight of polymers and the more important physical properties of compounded plastics.

Molecular Weight Determination

The significance of molecular weight, molecular weight spread and of the ratio between weight average and number average molecular weights of polymers has been described on pp. 28 and 29.

Elevation of Boiling Point
and Depression of Freezing Point of Solutions

These classical methods of molecular weight determination are, in theory, applicable to high polymers but, as the temperature changes becomes smaller with increasing molecular weight, use of the method is limited by the sensitivity of the measuring equipment. Even with ultra-sensitive equipment the maximum molecular weight that can be measured is about 40,000; the figure obtained is a number average.

Osmotic Pressure Measurement

This involves measuring the pressure difference developed between a dilute solution of the polymer and the pure solvent, separated by a semi-permeable membrane. The pressure difference is proportional to the number of molecules in solution and is, therefore, inversely proportional to the molecular weight of the solute. Molecular weights up to 10^6 may be measured in this way, a number average being obtained.

Viscosity Measurements

The changes in viscosity of polymer solutions with changes in concentration are related to the molecular weight of the polymer. The relationship is, however, complex and changes with the polymer and the solvent. The figure obtained is usually intermediate between weight and number average.

Viscosity measurements are usually made by determining the time taken for a fixed quantity of solvent and of polymer solution to pass through a capillary tube of known length and diameter; they require only simple apparatus, are easy to make and are, therefore, attractive as a control method. By determining the basic relationships between molecular weight (measured by an absolute method) and solution viscosity for a given solvent–polymer system, a procedure may be set up by which simple viscosity measurements of polymer solutions may be used as a routine method for control of molecular weight during polymer manufacture.

End-group Analysis

The number average molecular weight of linear polymers with distinctive end-groups, such as polyolefins, some polyesters or polyamides, may be calculated by determining the concentration of end-groups in a given weight of polymer; since this concentration decreases with increasing molecular weight, application of the method is limited by the sensitivity of the analytical procedure. For a branched chain polymer, when the molecular weight can be determined by some other method, end-group analysis will show the number of end-groups per molecule and, hence, the degree of branching.

Light Scattering Measurements in Polymer Solutions

This is one of the best ways of determining the weight average molecular weight. Very pure, dust free polymer and solvent must be used and both the turbidity and refractive index of the solution and solvent measured. Change in the refractive index is proportional to the number of polymer molecules in solution and to the amplitude of the vibrations set up. Increase in scattered light is proportional to the square of the amplitude, so measurement of both parameters enables the molecular weight to be calculated.

Sedimentation Methods

Polymer molecules in solution do not normally settle out under the influence of gravity but if a cell containing a polymer solution is rotated at high speed in an ultra-centrifuge, the polymer molecules tend to concentrate at the periphery. By passing a beam of light through the solution parallel to the axis of rotation, the change in refractive index and turbidity of the solution with increasing distance from the centre of rotation may be measured. Either the rate of sedimentation or the concentration gradient at equilibrium may be measured by this method and may be used to calculate the weight average molecular weight. The method requires elaborate and expensive apparatus.

MEASUREMENT OF PHYSICAL PROPERTIES OF PLASTICS

The measurement of physical properties for evaluation of plastics is especially important because the results of chemical and physical analysis cannot be related directly to mechanical behaviour. The tests described below employ mechanical methods and usually require the preparation of a test specimen, either by moulding or by other suitable means. The tests are mainly empirical and useful only for comparison and control purposes; for design calculations, much more extended tests over a wide range of conditions are required. Test results may be greatly affected by the previous history of the sample and by the conditions of test; the test procedure must, therefore, be carefully defined and rigidly adhered to if comparability is to be maintained. For most tests at least six duplicate determinations are required to give a representative result.

Most major industrial countries have laid down standard test methods; in the United Kingdom the standard methods of the British Standards Institution (B.S.I.) and of the American Society for Testing Materials (A.S.T.M.) are most commonly used.

Conditioning of Samples

This is an important procedure for ensuring reproducible results and involves holding the test specimens under standard conditions of humidity and temperature for an extended period, usually 40 hr, before the tests are carried out. The conditioning procedure is usually specified in the standard test method.

Tensile Strength

For this test a specimen of the shape shown in Fig. 13 is injection or compression moulded or cut from a $\frac{1}{8}$ in.-thick sheet.

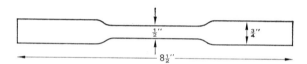

Fig. 13. Testpiece for tensile strength determination.

The tensile strength testing machine is provided with a pair of clamps, which can move apart at a predetermined rate, into which the specimen is fastened by its ends. When the machine is started the stress is automatically plotted against the strain so that not only can the stress necessary to break the sample, from which the tensile strength in lb/in² is calculated, be obtained but the modulus of elasticity and the elongation at break also become available.

Flexural Strength (Cross Breaking Strength)

A specimen 5 in. $\times$ $\frac{1}{2}$ in. $\times$ $\frac{1}{8}$ in. is prepared by methods similar to those used for obtaining the tensile test specimen described above. It is placed across two knife-edged supports 4 in. apart and a load, increasing at a specified rate, is applied to the centre as shown in Fig. 14.

The loading at failure is the flexural (cross breaking) strength and is expressed in lb/in². Most thermoplastics do not break under this test and the load needed to stretch the outer surface of the specimen by 5% is the figure taken.

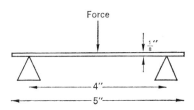

Fig. 14. Suspension of sample for determination of cross breaking strength.

Impact Strength

No really satisfactory method of determining the impact strength of plastics has yet been devised. The standard Izod method is the one most commonly used but, in spite of the most careful standardization, the results are not very consistent and only substantial differences between specimens are significant. The method depends on measuring the energy absorbed when a notched specimen is broken by being struck by a pendulum.

The specimen is usually 2 in. × ½ in. × ⅛ in. with a notch cut in its narrow face. The notched sample is clamped into the base of a pendulum testing machine with the notch facing the pendulum, which is held in a standard raised position. When all is ready the pendulum is released and strikes the specimen above the notch; the specimen breaks and the energy absorbed is calculated from the height reached by the pendulum on the follow through. The main function of the notch is to make sure that the specimen breaks in the same place on each repeat test. The results are expressed as ft-lb/in. of notch; this refers to the width of the notch and not the depth. In the standard specimen referred to the notch would be $\frac{1}{8}$ in. wide, so the figure obtained would be multiplied by 8. The test set-up is shown diagrammatically in Fig. 15.

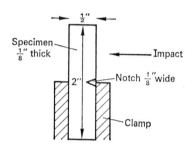

Fig. 15. Mounting of test specimen for Izod test.

Hardness (Rockwell)

This method depends on measuring the indentation when standard steel balls are applied to the surface of conditioned $\frac{1}{4}$ in. sheets, laid flat, under various loads. This test is applied particularly to rigid materials and the figures quoted refer to the size of steel ball and the loads; a knowledge of the test procedure is therefore required to interpret the results. Other methods may be used for flexible materials and for rubbers.

Specific Gravity and Density

For plastics these are normally determined at 23°C and are measured by the classical method of weighing the sample in water and in air. If the sample is a powder a weighed portion is added to a measured volume of water in a pyknometer and the volume change noted. Density is usually quoted in g/cm³. The difference between the figures for specific gravity and density is, of course, due to the fact that the density of water at 23°C is only 0·9976 g/ml.

Water Absorption

Discs of the material to be tested, 2 in. in diameter and $\frac{1}{8}$ in. thick, are dried in an oven for 24 hr at 50°C, cooled in a desiccator and weighed. The sample is then immediately immersed in water for a selected time, often 24 hr, at 23°C, wiped dry with a cloth and immediately weighed. The water absorption is reported as the percentage gain in weight.

This test is a useful indicator of electrical properties and dimensional stability; products giving high figures are often poor in both respects.

Dielectric Constant and Power Factor

Electrodes are applied to opposite faces of a preconditioned sheet of uniform thickness and the capacitance and dielectric loss measured by substitution methods in an alternating current electric bridge circuit.

The dielectric constant is the ratio of the capacity of a condenser having the plastics material as a dielectric to the capacity of the same condenser having air as the dielectric. It varies with the applied frequency.

The power factor is the ratio of the power loss in the plastics material when used as a dielectric to the total power passing through the system.

Dielectric Strength

Sheets of the material, previously conditioned, are placed between heavy cylindrical brass electrodes and an increasing voltage applied until the material breaks down. Breakdown means passage of a sudden excess current as shown by instruments or by visible damage to the specimen. Two tests are usually done:

1. The short time test in which the voltage is increased rapidly from zero to breakdown at, for example, 1 kV/sec.

2. The step by step test in which the initial voltage is 50% of the breakdown obtained in the first test and is then gradually increased.

The actual rates of increase are laid down specifically for each type of material to be tested. The results are expressed as volts per mil and, as the dielectric strength varies with the thickness, the actual thickness of the test piece must also be stated.

Deflection Temperature

This property was previously known as the heat distortion temperature and is the temperature at which a standard specimen undergoes a specified distortion.

A 5 in. $\times$ $\frac{1}{2}$ in. specimen, either $\frac{1}{8}$ in. or $\frac{1}{2}$ in. thick, is placed across supports 4 in. apart in a cabinet, the temperature of which can be raised at a controlled rate. A load to give a pressure of either 66 or 264 lb/in^2 is applied to the centre of the specimen and the temperature is raised at 2°C per minute. The deflection temperature (at 66 or 264 lb/in^2 respectively) is that at which the centre of the specimen has been deflected by 0·01 in. This test is useful for comparing samples but does not give a reliable indication of the suitability of any material for use at a given temperature.

Vicat Softening Point

This is the temperature at which a standard needle, area 1 mm^2, under a standard load, usually 1 kg, will penetrate the sample by 0·1 mm. The specimen should be at least $\frac{3}{8}$ in. square and $\frac{1}{8}$ in. thick. It is placed in a stirred bath at a temperature about 50°C below the expected softening point and the temperature raised at a rate of 50°C per hour. The needle is mounted vertically to bear on the sample; it is arranged so that weights may be placed on it and is attached to a gauge for indicating its movement. The temperature at which a penetration of the sample by 0·1 mm occurs is quoted as the Vicat Softening Point.

Flow Tests

The flow characteristics of plastics are of great importance in many process applications, especially in moulding and extrusion, but the development of a test which can be correlated with behaviour under operating conditions presents difficulties. A number of tests are in use but considerable experience is required in applying the results to practical problems.

For thermosets the cup flow test is widely used. It consists in moulding a standard cup under specified conditions, using an excess of moulding powder so that there is more flash than usual. The time is noted from the first application of pressure to cessation of flash movement.

For thermoplastics, injection moulding into a spiral mould under standard conditions can give a useful indication of the way a given material will behave in practice. A test applied specifically to polyolefins is determination of the melt flow index, illustrated in Fig. 16.

The sample is contained in an open topped cylinder with the jet, $0 \cdot 315$ in. long and $0 \cdot 0825$ in. in diameter at the lower end.

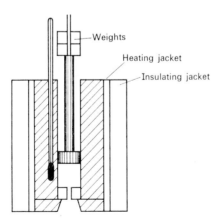

FIG. 16. Diagram of apparatus for determination of melt flow index.

P.S.R.—C

A piston, on which variable loads may be placed, slides into the cylinder on top of the sample and the whole apparatus is maintained at a standard temperature, 190°C for polyethylene and 230°C for polypropylene. Four to five grams of sample are placed in the cylinder and the load, 2·16 or 5 kg according to the flow properties of the sample under test, is applied. The molten plastic is extruded through the jet and, at suitable intervals of time, the extrudate is cut off and weighed.

The melt flow index is:

$$\frac{600 \times \text{average weight of cut off in grams}}{\text{time interval between cut off in seconds}}.$$

READING LIST

Physical Properties of Plastics, by Bueche, Interscience, New York, 1962.
The Properties and Testing of Plastics Materials, by A. E. Lever and J. Rhys, Temple Press, London, 1968.

PART II

Thermosetting Resins

CHAPTER 3

Phenolic Resins and Moulding Powders

PHENOLIC resins are among the oldest known synthetic resins. Many workers were studying the reactions between phenols and aldehydes towards the end of the nineteenth century and the production of a resin from them was noted as early as 1872. The commercial possibilities of these products were not realized until the early years of the twentieth century when the so-called shellac substitutes were produced. Baekeland in the United States worked in this field for 5 years before he filed patents covering the basic principle of making fusible resins which could be converted to the infusible state in the mould in the presence of a hardening agent. James (later Sir James) Swinburne worked along somewhat parallel lines in the United Kingdom but was narrowly anticipated by Baekeland in applying for a patent. He continued his work, however, mainly in the synthetic lacquer field and set up the Damard Lacquer Company in 1910. Meanwhile Baekeland had established the Bakelite Company in the United States but made slow progress due to the difficulty of persuading users to adopt the new techniques. The developing motor industry, however, found uses for his products in distributor heads, switches and junction boxes and this started the use of Bakelite resins as electrical insulators.

During the next 10 years other uses for these resins were developed, for example in the production of gramophone records, adhesives and varnishes, and in 1922 many of the rival interests in the United States were merged into the Bakelite Corporation. Meanwhile Swinburne had arranged to use the Bakelite patents and rapid development took place in the United Kingdom. In

1928 Bakelite Limited, the British subsidiary of the Bakelite Corporation, took over the interests of the Damard Lacquer Co. and Swinburne was made the first chairman of the new company. In 1940 the Bakelite Corporation in the United States was taken over by the Union Carbide Corporation, but Bakelite Limited still maintained a semi-independent existence until 1962 when Union Carbide and the British firm, the Distillers Company Limited, merged their U.K. plastics interests in Bakelite Xylonite Ltd. and Bakelite's remaining independent shareholders were bought out. The Bakelite Group had, by this time, established themselves in many other countries and, as the original Bakelite patents had long since expired, competitive manufacture was established all over the world. Baekeland himself lived until 1944 and Swinburne until 1957, when he died at the ripe old age of 99.

The history of the Bakelite Group has been given in some detail because it was at the heart of the first big development of the modern plastics industry. Its trade mark "Bakelite" became a household word in the years after the 1914–18 war and was applied indiscriminately by the man in the street to any material which looked as if it might be made of synthetic resin. Production of phenolic resins is still increasing, albeit somewhat spasmodically. Several times it has appeared that growth had stopped and then, a year or two later, a new surge in output has started the upward trend again. Table 1 shows the growth in

TABLE 1. PRODUCTION OF PHENOLIC RESINS IN
THE UNITED KINGDOM
(thousands of long tons)

1954	42·5	1961	55·7
1955	45·9	1962	56·3
1956	45·9	1963	57·7
1957	48·2	1964	64·0
1958	47·0	1965	66·2
1959	54·2	1966	65·6
1960	58·8	1967	63·6

output over the past 14 years. It covers phenolic resins for all applications and includes the water content of water soluble resins (see p. 73).

CHEMISTRY OF PHENOLIC RESINS

The phenolic constituent of the resins may be phenols, cresols, xylenols, dihydroxy phenols or substituted phenols, while the aldehyde can be formaldehyde, acetaldehyde or furfuraldehyde; by far the major proportion of the resins of commerce are made from phenol or cresols with formaldehyde. Phenol has three points reactive to formaldehyde in the molecule, two in the ortho- and one in the para-position with respect to the hydroxyl group; it is, therefore, a trifunctional compound and in conjunction with the difunctional formaldehyde can form cross-linked structures.

The first stage of the reaction is very simple and consists in the addition of formaldehyde at one of the reactive points to give a phenyl alcohol of the type $C_6H_4\begin{subarray}{l} \diagup OH \\ \diagdown CH_2OH. \end{subarray}$ From this point the reaction becomes more complex and its course depends on the concentration of the reactants and the conditions. The methylol ($-CH_2OH$) groups will react further, either with each other or with one of the active points in the phenol molecule, with elimination of water, to build up a chain of phenol molecules connected by methylene or methylene ether linkages. At the same time free formaldehyde, if present, will add on at other reactive points in the phenol molecules to give additional reactive methylol groups which can then react further to cross-link the chains. The first stage of the reaction proceeds very rapidly but the later stages are much slower and are promoted by catalysts of various kinds. In practice two distinct types of resin, commonly called novolaks and resols, are produced according to the reaction conditions.

Novolaks

These are resins which have no residual reactive methylol groups and are incapable of cross-linking without the addition of a hardening agent. They are made by condensing approximately equimolar proportions of the phenol and aldehyde under acid conditions; a typical novolak chain from phenol and formaldehyde is shown below:

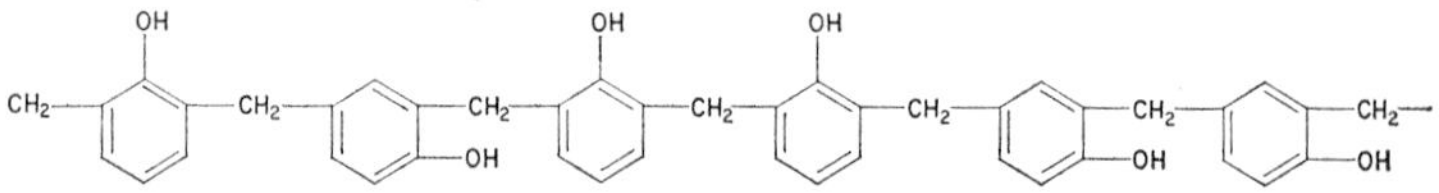

In practice this is a brownish yellow, hard and brittle resin usually containing traces of unreacted phenol. When heated with additional formaldehyde, usually supplied in the form of its crystalline reaction product with ammonia, hexamethylene tetramine (hexa), the resin chains cross-link to form infusible resins insoluble in any solvents. A typical scheme is shown.

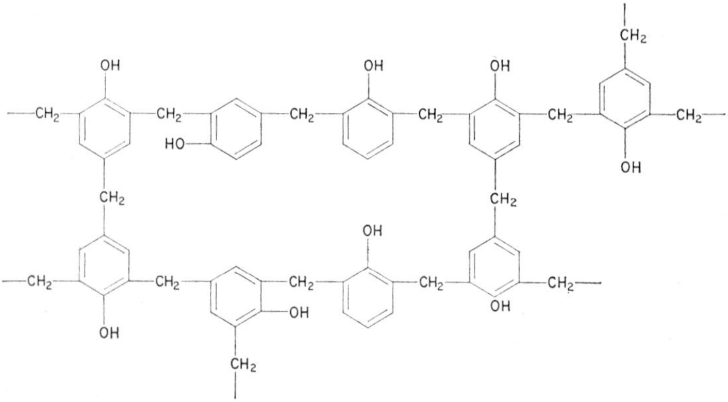

In practice this reaction is made use of in the production of moulding powders. The novolak resin can be ground, blended with fillers, colouring matter and hexa and cured to the infusible state in a hot mould. Water is evolved during the curing process, which is described in more detail on p. 70.

Resols

These resins are made by condensing phenol with an excess of formaldehyde under alkaline conditions. The first chains formed have free reactive methylol groups available and conditions are so arranged as to stop the reaction at this stage.

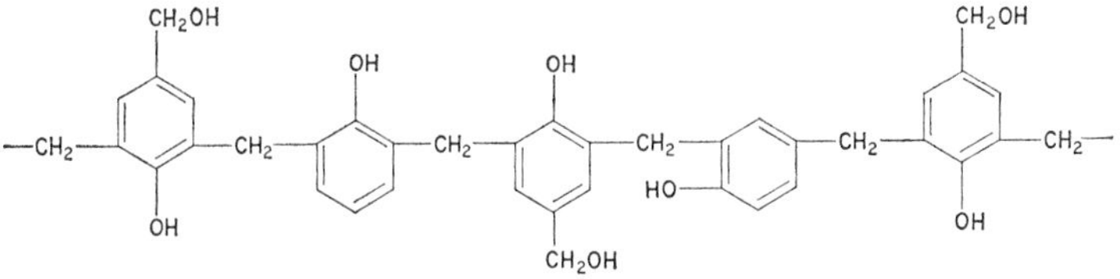

The CH_2OH groups can cross-link adjacent chains with elimination of water without the aid of additional formaldehyde. If ammonia is used as the alkaline catalyst it has been shown that benzylamine groups and, later, some cross-links containing NH groups can be formed in the resin. A possible method is shown below:

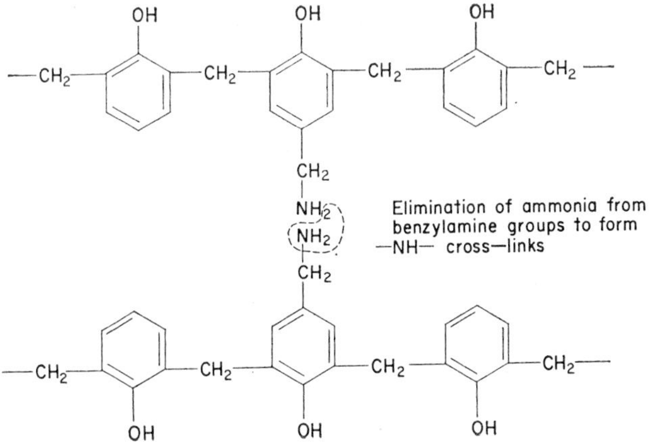

Elimination of ammonia from benzylamine groups to form —NH— cross—links

Resols are normally used for impregnating paper or textiles in the production of laminates for structural or decorative purposes; the impregnated sheets may be cured without further chemical addition by the application of heat and pressure alone.

The reaction schemes described above, both for resols and novolaks, have been greatly over-simplified; the actual reactions occurring in commercial manufacture are certainly much more complex, especially when cresols or other substituted phenols are used.

RAW MATERIALS

Phenol was originally produced by distillation of coal tar and some 10,000–15,000 t/a are still made in this country from that source. Output of synthetic phenol has steadily increased over a period of many years and now makes up the major part of U.K. production, which totals nearly 100,000 t/a. The favoured process is through air oxidation of isopropyl benzene (cumene) to the hydroperoxide which is then rearranged and hydrolysed by acid to form phenol and acetone. Phenol is a solid melting at 41°C, so that jacketed and insulated equipment is needed for its handling and storage. It may also be rendered liquid at ordinary temperatures by the addition of 10% of cresol and, in the distillation of coal tar, this mixture may be directly produced as a fraction by suitable control of distillation conditions.

Cresols

Coal tar is still virtually the only source of cresols; as produced the commercial products are mixtures of the three possible isomers. Only the meta isomer is trifunctional, and thus capable of giving cross-linking resins, since the other isomers have one of the ortho or the para positions with respect to the hydroxyl group already occupied by the methyl group. Meta cresol is never used alone, however, both on account of its cost and because its reaction with formaldehyde is uncontrollably violent. Careful fractionation of coal tar acids will give a fraction containing mainly ortho cresol and one containing varying proportions of meta together with para and some ortho, and these meta-containing fractions are used in commercial resin manufacture. Standard grades on the market contain 41–2%, 49–50% and 52–3% respectively. Grades with higher meta content are available for special purposes.

Cresol may be made synthetically from toluene, and some relatively small quantities are being produced commercially. Production of the high proportions of the meta-isomer required for plastics manufacture at an economic cost has presented some problems and the synthetic material is little used for cresol–formaldehyde resins.

Xylenols

These too are obtained from coal tar and six possible isomers exist. Only the 3–5 isomer reacts as a trifunctional compound and its reaction with formaldehyde is even more violent than that of meta cresol; the other isomers will react but rather slowly. The production of fractions rich in the 3–5 isomer is possible by careful distillation of coal tar but, because of the number of isomers present, including also ethyl phenols, the products are very difficult to standardize. Thus different batches of a xylenol fraction of defined 3–5 content may vary greatly in reactivity and this, in turn, makes the production of resins of constant quality almost impossible.

Other substituted phenols

Para-phenyl phenol, *p*-tertiary butyl phenol and diphenylol propane are all used for special applications, mainly in the production of surface coating resins, but the quantities are not large. Chlorinated phenols have been suggested for production of fire resisting resins.

ALDEHYDES

Formaldehyde

All the formaldehyde of commerce is made by the catalytic oxidation of methanol with air and U.K. output now amounts to over 80,000 t/a. The compound is a gas at normal temperatures and is generally used as an aqueous solution of about 40% strength. Alternatively a solid polymer, "paraform", may be used. This may be obtained by vacuum dehydration of a formaldehyde solution and consists of a mixture of $HO(CH_2O)_nH$ compounds; it is more convenient to handle but more expensive. These polymers are also formed in aqueous solutions of formaldehyde and may precipitate at room temperature. It is essential in industrial practice to prevent this precipitation and formaldehyde must be stored at about 50°C; alternatively 5–6% of methanol added to the aqueous formaldehyde is sufficient to maintain a homogeneous solution and has the further advantage that oxidation to formic acid is almost prevented. It is possible to produce such a solution by control of conditions during the methanol oxidation process. Methanol-free formaldehyde has, however, some advantages for resin production since it reacts more rapidly and produces a lighter coloured resin.

Acetaldehyde

Acetaldehyde has been made synthetically by the controlled oxidation of ethanol for many years. A more recent development, which seems likely to replace the older method, is the Wacker process for direct oxidation of ethylene in the presence of palladium and copper compounds. It is most conveniently used as its trimer, paraldehyde, and will react with phenols to form resins of the novolak type; resols cannot, however, be formed. The novolaks are dark coloured, soluble in drying oils and can be cured slowly with hexa. Their oil solubility makes it possible to produce a moulding powder curing to slightly flexible products, which has found limited uses, but they have not developed commercially to any great extent.

Furfural

This cyclic aldehyde is at present made from oat hulls by hydrolysis of the pentosans contained in them to pentoses; these then split off water to form furfural. No commercial synthesis has been developed but there are vast quantities of vegetable waste such as sugar cane bagasse, corn cobs and waste liquor from wood pulp processes from which furfural might be recovered if required. It has been used fairly extensively in the United States in the production of phenol–furfural resins; the reaction is very complicated as the furfural, in addition to reacting with the phenol, can itself resinify. Novolaks are best made with a small proportion of furfural (20%) on the phenol; they can be cross-linked with hexa, curing slowly at 150°C or more rapidly at 200°C. Resols can be made but are little used. Useful properties of phenol–furfural resins are the ease with which water may be removed, due to the high boiling point of furfural, and the fact that the resins are soluble in excess furfural making it easier to obtain homogeneous products.

HARDENERS

Hexamethylene Tetramine

This compound, commonly known as hexamine or hexa, is virtually the only hardening agent in large scale use for phenol–formaldehyde resins. It is made by reacting aqueous formaldehyde with ammonia and crystallizes out of the resulting solution.

$$6CH_2O + 4NH_3 \rightarrow N_4(CH_2)_6 + 6H_2O$$

MANUFACTURE OF PHENOLIC RESINS,
MOULDING POWDERS AND LAMINATES

The reactions leading to the formation of phenolic resins are condensation reactions involving the splitting off of water during the condensation process and the subsequent removal of this water from the reaction mixture. These operations are carried out in a resin kettle and this piece of equipment, as described below, may also be used virtually unchanged for the manufacture of amine and polyester resins.

The kettle is usually a cylindrical vessel with a curved bottom and a flange at the top to which a cover is bolted. It is provided with a flanged opening at the bottom to which a discharge valve is bolted and a jacket to which steam or water may be admitted for heating and cooling. Important points of construction are that the discharge valve must be wide enough to permit the rapid discharge of the batch of viscous resin and that the jacket should completely surround the discharge valve opening to prevent the creation of a cold pocket in which highly polymerized resin could solidify and prevent the discharge of the kettle contents at the end of the reaction. The cover carries the driving gear for the stirrer, which passes through its centre and reaches to the bottom

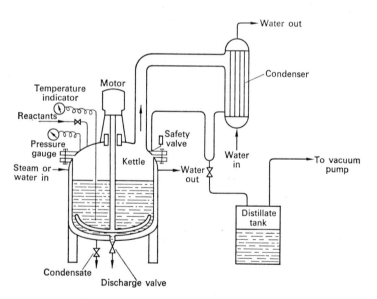

Fig. 17. Diagrammatic arrangement of typical resin kettle.

of the kettle. It is important that there should be only a small clearance between the stirrer blades and the side and bottom of the kettle to prevent the formation of an overheated film of resin where excessive polymerization could take place. The cover is also provided with openings for a vapour pipe, which leads to a condenser, and for temperature and pressure measuring instruments; it is often fitted with glass portholes through which the contents may be inspected during the reaction. The condenser is used for providing reflux to the kettle and for distilling off water and excess reactants from the kettle. It is provided with a distillate receiver and is connected to a vacuum pump for distillation under reduced pressure. Mild steel is the preferred metal of construction but, when reactants are corrosive and colour of the product is important, nickel, copper, stainless steel or glass-lined equipment may be used according to circumstances. A diagram of a typical resin kettle is shown in Fig. 17.

Novolak Resins and Moulding Powders

An average moulding powder novolak might be made by mixing 100 parts by weight of phenol, 70 parts by weight of aqueous formaldehyde (40% w/v) and 0·3 parts by weight of hydrogen chloride. The exact amount of acid is controlled by a pH determination as too low a pH may lead to excessively violent reaction. The mixture is stirred vigorously and heated gradually until reaction starts; once started the exothermic reaction is controlled by cooling in its early stages. As the reaction rate slows down the heat is turned on again and the mixture refluxed for several hours, when it separates into two layers; the degree of polymerization is determined by measurement of some physical property of the resin, such as the refractive index, which changes as condensation proceeds. When the desired end point has been reached, the excess water is distilled off under vacuum until a sample of the resin, cooled quickly, breaks cleanly and is no longer cloudy. The resin is then run off from the base of the kettle into metal trays, allowed to solidify and ground to a fine powder. At this stage the resin is tested for free phenol content, gelation time (with 10% added hexamine) and flow. Too high a phenol content will cause soft mouldings. Gelation time and flow are important properties determining the behaviour of the resin during the moulding process and give a measure of the amount of added hexamine and heat treatment which will be required during the conversion of the resin to a moulding powder.

The moulding powder can now be formulated; a typical formulation might be 100 parts of novolak resin, 12 parts of hexa, 100 parts of wood flour (as filler), 2 parts of stearic acid (as mould lubricant), 4 parts of pigment and 4 parts of magnesium oxide. The magnesium oxide serves to neutralize any traces of acid from the original catalyst, which might be evolved during the cross-linking process. The ingredients are thoroughly mixed in a mechanical mixer and then transferred to heated roll mills. The rolls are mounted horizontally and parallel, and are hollow so that they may be heated by steam passed in through the end bearings; they revolve in opposite directions and at different speeds and the distance between them (the nip) is adjustable. The front roll is kept at about 85°C and the back one about 110°C. In contact with the hot rolls the resin melts and binds the other constituents together while the differential speed mixes the material as it is formed into a soft thin sheet; the high temperature causes the resin to react with the hexamine and the mixture starts to "cure". The sheet is continually cut away and repassed through the rolls until the operator judges that the cure has advanced to the optimum stage; the aim is to produce a material which will complete its cure quickly in the mould and yet still flow freely enough to fill all parts of the mould. This depends largely on the skill and judgement of the operator. When ready, the sheet is cut from the rolls, cooled, crushed and ground to the finished moulding powder. Normally, several "sheetings" from the rolls are blended together to increase uniformity. Finished moulding powders are usually tested by actually making test mouldings and examining their characteristics and properties.

Effect of Varying the Constituents

The example given would produce a low priced, general purpose moulding powder. At one time cresol would have been used to cheapen the product but phenol is now cheaper than cresol so there is no point in the change. Cresol resins give lower strength mouldings but are rather more acid resistant. When using cresols the ratio of the isomers must be carefully controlled to ensure uniformity; the higher the meta content the faster the moulding powder cures. Xylenols are used only in special cases where alkali resistant resins are required. Colour is improved by the use of methanol-free formaldehyde and of oxalic acid instead of hydrochloric acid as a catalyst.

The wood flour filler in the example gives a relatively low strength moulding; higher strength can be obtained by using shredded cloth as a filler, the best heat resistance by using asbestos fibre and the best electrical properties by using mica. The general method of production is similar in all cases except that when using shredded cloth, which is difficult to mix with the other ingredients, a bladed mixer of the Banbury type may be used to give a thorough premix.

Producers and Users of Phenolic Moulding Powder

The production of phenolic moulding powders in the United Kingdom over the past few years has been roughly constant as shown in Table 2.

TABLE 2. U.K. PRODUCTION OF PHENOLIC MOULDING POWDERS
(all figures in 000 tons)

1962	29·7
1963	30·8
1964	35·4
1965	34·7
1966	31·8
1967	30·4

Their applications have been divided between the industries listed in Table 3:

TABLE 3. DISTRIBUTION OF
PHENOLIC MOULDING POWDERS BETWEEN INDUSTRIES

	%
Electrical uses	35
Household goods	15
Motor-cars	10
Sanitary equipment	10
Closures	10
Furnishing	8
Miscellaneous	7
Textile industry	5

Since the production of phenolic moulding powders does not require excessively large capital investment, many small firms have been encouraged, at various times, to undertake manufacture. There has been a good deal of concentration of the industry in the years since the Second World War and, although a number of smaller producers still flourish, the major part of U.K. production is divided between four companies, namely B.X.L. Plastics Materials Group Ltd., B.P. Chemicals (U.K.) Ltd., Sterling Moulded Materials Ltd., and James Ferguson & Co. Ltd.

RESOLS AND LAMINATED PLASTICS

Laminates are not restricted to phenolics and other resins, especially amino resins, may be used; this is, however, a major outlet for resols and a general description of laminated plastics has been included in this chapter.

The Resins

A two-stage curing process, such as is used in the production of moulding powders, would be difficult, if not impossible, to apply to the production of laminates. A resol, which will cure without further chemical additions, is therefore usually used as the impregnating resin.

The equipment used for resol manufacture is similar to that for novolaks described on p. 68. One hundred parts by weight of phenol, 100 parts by weight of 40% w/v formaldehyde and nine parts by weight of 25% ammonia are mixed in the kettle and the temperature is raised cautiously to boiling point. The mixture is refluxed, with constant stirring, until the reaction has proceeded to the required degree as shown by, for example, the refractive index. Vacuum is now applied and water and excess formaldehyde distilled off at a pressure of about 100 mm, so that the temperature of the resin is kept well below 100°C. Stirring is continued during distillation, which is carried on until a sample of the resin, drawn off through a vacuum trap, cools to give a solid which breaks crisply. The resin may now be discharged into trays, broken up, and ground; but more frequently a solvent, such as industrial methylated spirit, isopropyl alcohol or acetone, is added to the kettle as soon as the vacuum is broken to produce a resin solution containing 50–60% resin.

Manufacture of resols needs careful control; if the proportions of reactants are not quite right or the temperature rises too high, or remains high too long, the resin may cure in the kettle. The curing takes place very quickly; once it starts in any part of the resin mass it spreads in a few seconds to the whole batch and there is nothing to do but to wait until the mass has cooled and chip it out from inside the kettle.

If caustic soda ($0 \cdot 2$–$0 \cdot 6\%$ on the phenol) is used instead of ammonia a resin of the required degree of condensation can be obtained which is still homogeneous and which may even be diluted with water. Such resins cure faster than the ammonia catalysed ones, are cheaper since no solvent is needed, but are not suitable for electrical applications.

Cresol resins are often preferred to phenol for laminates; a commercial cresol with 42% of the meta isomer gives a resin which cures reasonably fast but for which the curing conditions are less critical than for phenol resins. This means that there is a much greater margin of safety at the baking stage. Homogeneous resins cannot be prepared by using caustic soda as a catalyst and a solvent is almost always necessary. Other things being equal, cresol resins give marginally better electrical properties than phenol resins.

As in the case of novolaks, xylenol resins are reserved for those applications where especially high resistance to alkalis is required.

Plastic Laminates

Laminated plastics are essentially layers of paper or cloth impregnated with a resin and bonded together by heat and pressure. Sheets, rods and tubes are most commonly produced; more intricate shapes can be made but the moulds are expensive. The distinction between a reinforced plastic and a laminated plastic is becoming very blurred; as described under unsaturated polyesters on p. 97, a reinforced plastic can be formed into shape in the simplest possible mould from glass fibre and an unsaturated polyester resin. Pressure is not usually required and a cold cure can be effective. Gradually, however, the value of pressure in producing articles of reinforced plastic has been realized and, moreover, sheets of glass fibre can now be bonded with other resins, including phenolic, as well as with unsaturated polyesters. Originally the conventional laminating resins were cured under pressure that damaged glass fibre and this made a clear distinction between laminates and reinforced plastics; more recently phenolic resins have been produced which can be cured at relatively low pressures and are suitable for use with glass fibre so that the distinction is tending to disappear, with a resulting confusion in nomenclature in the fringe products.

As well as the phenol, cresol and xylenol based resins already mentioned, urea and melamine resins are also used, the last-named usually for the surface layer only on account of its high cost.

The Filler

The sheet material may be paper, including asbestos paper, or almost any woven material that can be produced in sheet form. Cellulose paper can be rag paper, made from pure cotton fibres, which gives the highest strength laminates, paper from purified wood pulp giving exceptional electrical properties, or kraft paper giving the cheapest product.

Textile fibres give higher strength to the finished laminate than paper although they are not generally so good for electrical purposes. Plain woven cotton fabrics varying in weight from 2 to 10 oz/yd^2 are mainly used; the heavier weights give greater strength while the lighter cloth produces a material of better machinability. Synthetic fibre cloth, especially nylon, may be used but price has, in the past, been an inhibiting factor.

Asbestos cloth is used on a limited scale for heat resistant products but gives a lower strength unless a very high grade and expensive asbestos cloth is used. A cheaper product with good heat resistance and high strength can be made from asbestos flock. Asbestos based products are often used for bearings, gears and other engineering products as they have a low coefficient of friction when lubricated with water and are very durable.

Impregnation

The paper or fabric is dried to a low moisture content; it is then passed over guide rollers 36–43 in. wide, through a trough of resin solution and then through squeeze rollers or under a doctor blade to remove excess resin. Lengths of several hundred feet are treated at a time. The quantity of resin picked up by the sheet must be controlled; it varies with the viscosity of the resin solution and with the degree of squeeze after impregnation. The viscosity of the resin solution in the impregnating tank is maintained constant by adding solvent as required and is generally controlled automatically. For most products a resin pick up of 30%, calculated as dry resin on the paper or cloth, gives the best results although higher resin pick ups are advantageous for some bearings where a low coefficient of friction is more important than mechanical strength.

The Drying Stage

After leaving the doctor blade or squeeze rolls the impregnated materials pass through a long drying chamber, where the volatiles are removed under carefully controlled temperature conditions, and are then cooled and wound on to mandrels until required for laminating. Impregnated paper is often sold in this condition for the preparation of laminates by other manufacturers. The exact condition of dryness is critical; to produce a sheet with the best electrical properties it is necessary to take the drying stage to the limit, leaving only just sufficient flow in the resin to permit the layers to bond when the laminate is heated under pressure. As noted on p. 73, the drying stage is less critical with cresol resins due to residual thermo-plasticity of the ortho- and para-cresol–formaldehyde resins. Resins dissolved in low boiling organic solvents give better results for electrical purposes than water soluble resins since, with the latter, the desired degree of cure of the resin may be reached before all the water has been removed from the sheet. Water soluble resins will, however, give a sheet with excellent mechanical properties under the proper conditions.

Pressing

Pieces of impregnated material are plied up to give the desired thickness of finished sheet and the pack placed between metal sheets, chromium plated if a good surface finish is required. Several of these assemblies are loaded into a multiplaten press and pressed together until the resin is cured. The platens are heated by steam or electricity to about 150°C but the temperature can vary according to the type of resin being used. The pressure is usually between 1000 and 2000 lb/in² and a time of 30 min is usually adequate. For maximum flatness the sheets are allowed to cool slowly in a warm press.

When decorative laminates are being prepared, it is usual for the top and bottom sheets to be impregnated with a melamine formaldehyde resin and, where a pattern is required, the last sheet is a printed paper. The cured sheets may be ground, bandsawn, drilled, punched or machined in other ways as required. Sometimes the sheets are ground to remove the surface layer of resin; closer thickness tolerances may be obtained in this way but the appearance of the surface is completely changed. Sheet based on asbestos cloth is frequently treated in this way as it is difficult to get a really good appearance from the platen surface, while the coarse asbestos cloth makes thickness control difficult.

Production of Tubes and Rods

Tubes are frequently made from impregnated paper by winding the paper on to a mandrel and curing in an oven. Provided a suitable resin is used this can give a satisfactory product and is the normal way of making electrical bushings. Fabric bearings and gears are made by winding the fabric on a mandrel but tension is more difficult to apply, so that the wrapped mandrel must be cured in a suitable mould (made in two halves) in a press. Large gears are usually cut from the correct thickness of sheet and machined to size; various sections can be made by using the appropriately shaped moulds. Rods are made by winding the impregnated cloth or paper on a very thin mandrel, removing the mandrel and immediately pressing in a mould to cure.

The advent of satisfactory unsaturated polyesters which will cure without pressure or at very low pressures has limited developments in the application of phenolic resin laminates to special shapes.

Production

The total U.K. production over the past 6 years has been as in Table 4.

TABLE 4. U.K. PRODUCTION OF PHENOLIC RESIN LAMINATES
(all figures in 000 tons)

	Decorative sheet	Industrial sheet	Tubes, rods, etc.	Total
1962	11·9	10·2	1·2	23·3
1963	12·1	10·9	1·3	24·3
1964	13·4	11·8	2·0	27·2
1965	14·2	11·9	1·5	27·6
1966	16·2	10·6	1·4	28·2
1967	16·6	10·0	1·2	27·8

The principal producers of laminated plastics are: B.X.L. Plastics Material Group Ltd., Formica Ltd., Tufnol Ltd., The Bushing Co. Ltd., Micanite & Insulators Ltd., J. W. Roberts Ltd., Commercial Plastics Ltd., Permali Ltd.

SURFACE COATING RESINS

While the alkyds, described on p. 99, have for many years been the most widely used synthetic resins for paint manufacture, phenolics have been found very valuable for certain purposes. Their use dates back to the days of the First World War when Dr. Kurt Albert in Germany worked out techniques for modifying phenol resins to make them compatible with paint and varnish constituents and marketed his products under the trade name "Albertol".

The most widely used method is to add rosin (largely abietic acid) to a partly reacted, acid catalysed phenol–formaldehyde mixture and then to heat to about 200°C to complete the reaction with the rosin. Alternatively, resols can first be formed and then heated with rosin to temperatures of 200–250°C. Both methods have for their purpose the production of a resin which will dissolve in drying oils to give paints or varnishes of better performance than those made with rosin alone. These paints will dry faster and give harder and more water-resistant coatings, while the good wetting and dispersing properties of the rosin are not lost. The chemical reactions involved in the formation of the resin are very complex but it is known that methylol phenols will add to the olefinic linkages present in abietic and similar acids. The drying oils commonly used are linseed and tung oils.

Although the simple phenol–formaldehyde resins are not compatible with drying oils, cresol resins have some compatibility and the presence of bulky side chains on the aromatic nucleus confers much greater compatibility. The so-called 100% phenolics are made by condensing, for example, *p*-tertiary butyl phenol with formaldehyde using an acid catalyst. Resins of this type are thermoplastic since only the two reactive positions ortho to the hydroxyl group remain free so that the phenol becomes, in effect, a difunctional compound. They can, therefore, be prepared as hard solids which will dissolve in drying oils by heating to high temperature with stirring; paint formulated from these resin solutions air dries in the usual way with the aid of the driers included in the formulation. These paints have high water and chemical resistance and are particularly useful in marine applications. These resins, however, are not cheap and paints based on epoxy resins are, to some extent, taking their place as they have other advantages such as hardness and outstanding adhesion to metal. All paints based on phenolic resins tend to turn yellow on standing, especially on exposure to strong light, so that they are little used for decorative purposes.

USE OF PHENOLIC RESINS IN METALLURGY

Aqueous emulsions of liquid resin made from cresol and formaldehyde are used for making cores for foundry work. Sand is mixed with 1–2% of the aqueous resin emulsion, together with a little dextrin, and the mixture forced into core boxes to form a core. The core is then carefully removed from the box and cured in an oven. The cured core is strong and may be used for casting any metals normally cast in sand. Shell moulds are made in a somewhat similar way; usually a phenol novolak is used with 10% of hexa and some 5% of this mixture is blended with sand. The metal pattern, provided with ejector pins, is heated to about 200°C and the resin mixture is poured over it. After 30 sec, or longer according to the thickness of mould required, excess resin powder is tipped off and the metal pattern with its adherent resin transferred to a curing oven. After cure the mould is removed from the pattern with the aid of the ejector pins. The process is illustrated in Plate II. These moulds are excellent for metal castings and have many advantages; they have good dimensional stability and give products with better surface finish and fewer faults than moulds made by other means. They are also relatively light to stock and have a long life.

USES OF PHENOLIC RESINS AS ADHESIVES

Laminating resins, as already described, are really hot setting adhesives and may be used as such, especially in the manufacture of plywood. If a strong acid is added to a liquid resol in an advanced stage of condensation, the resin cures in a few seconds with evolution of heat and much foaming and bubbling is caused by the water of condensation. By choosing a suitable acid, such as *p*-toluene sulphonic acid, the reaction can be controlled and such mixtures can form the basis of cold setting adhesives. Fillers may be incorporated so that gap-filling cements can be produced. The residual acidity is, however, a disadvantage and makes them unsuitable for use with metals to which, in any case, they do not adhere well. A mixture of a resol with polyvinyl formal (see p. 224) will bond with metal surfaces but must be hot cured. The polyvinyl formal may be mixed with the resol in a mutual solvent, or may be applied as a powder to one of the faces to be joined while the phenolic resin is painted on to the other. On

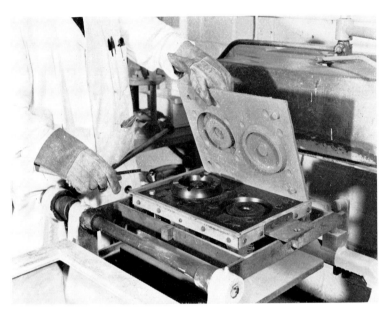

PLATE II. A shell mould being prepared. (By courtesy of British
Resin Products Ltd.)

bringing the faces together and heating, a good bond will be formed.

Adhesives based on phenol and cresol are resistant to boiling water; they withstand weak acids and alkalis but they are brittle and dark coloured. Much better cold setting adhesives may be made from resins based on resorcinol (meta-dihydroxy benzene). Resorcinol reacts very rapidly with formaldehyde and the resin is prepared by adding the formaldehyde gradually to the resorcinol, with constant agitation, so that there is never an excess of formaldehyde in any part of the mixture; in this way a stable product with a reasonable shelf life may be produced. These resins may be made to set in the cold under virtually neutral conditions by the addition of formaldehyde in the form of its polymer, paraform. Resorcinol-based adhesives are used as the adhesive for marine plywood; they would be much more widely used if they were cheaper but resorcinol is still made by disulphonation of benzene, followed by fusion with caustic soda and extraction with ether, from which the resorcinol is recovered. The present U.K. demand of 500–600 t/a does not justify a more sophisticated method but, until the cost is reduced, demand will not greatly increase.

Composite adhesives made from phenol–formaldehyde resins and synthetic rubber are of considerable importance. The synthetic rubber component may be of the "Neoprene" (polychloroprene) or nitrile type. Neoprene itself in solution is a useful cold setting adhesive but the bond is greatly strengthened if the neoprene is compounded with a phenolic resin and hot cured. A short chain neoprene polymer is compatible with resins based on *p*-tertiary butyl phenol or with terpene modified phenolic resins. The blend might consist of 100 parts of neoprene, up to 50 parts of phenolic resin and 50 parts of fillers together with small quantities of magnesium oxide, zinc oxide and an accelerator which does not react with phenolic resins such as thiocarbanilide. This adhesive can be used for bonding to metals; it has very good bond strength while resistance to impact and fatigue are also good. It also stands up well to oils, aircraft fuels and chemicals generally. Normally hot cured, it may be formulated for cold-setting but, in this case, it is doubtful if the full effect of the phenolic resin is obtained.

Nitrile rubber, a copolymer of acrylonitrile and butadiene, can be blended with phenolic resins by milling or by dissolving in a mutual solvent such as methyl ethyl ketone. The normal rubber curing agents must be added and sometimes adhesives of this kind are marketed as two-component systems. Hot curing is the normal practice. Applications include bonding metals, especially light alloys, copper to plastic laminates for printed circuits, and asbestos–resin brake linings and clutch discs to metallic brake shoes and clutch plates for the automobile industry.

FRICTION MATERIALS

Brake linings and clutch discs can be made from phenolic resins. In the production of the former, multilayer asbestos cloth is impregnated with a varnish made from a medium meta content cresol–formaldehyde resin condensed with excess ammonia, the water largely run off and the resin heated with a drying oil, such as china wood oil, until completely soluble in solvent naphtha or xylene; the viscosity of the varnish is controlled by the amount of solvent added. The impregnated asbestos is dried, compressed and formed to shape and finally baked. Clutch discs may be made using a similar impregnant; alternatively a straight cresol–formaldehyde resin catalysed with caustic soda gives good results. Single or multiply cloth may be used as required.

This type of friction material has been largely replaced by powder mixes, so called moulded linings or discs, in which asbestos fibre, fillers and a phenol–formaldehyde resin are formed into sheets, partly cured and shaped and baked to complete the cure. The resin is an acid-catalysed solid resin with a high proportion of formaldehyde; hexamine is added at the blending stage. Many of the pads for the modern disc brakes contain a special phenol–formaldehyde resin formulated to give maximum heat resistance.

OTHER USES FOR PHENOLIC RESINS

There are many other relatively small scale applications for phenolic resins of which a few are listed below:

(i) Small items of chemical plant.
(ii) Cast products—not now very important.
(iii) Expanded foam—now largely replaced by other resins.
(iv) Grinding wheels and abrasive discs.
(v) Densified wood.
(vi) Electrical coil impregnation.

In general, phenolic resins may be regarded as materials which are coming to the end of their possibilities for development but which are likely to remain of industrial importance and to be produced in substantial quantities for many years to come.

READING LIST

Phenolic Resins, by A. A. K. Whitehouse, E. G. K. Pritchett and G. Barnett, Iliffe Press for the Plastics Institute, London, 1967.
The Chemistry of Phenolic Resins, by R. W. Martin, John Wiley, New York, 1956.
Phenolic Resin Chemistry, by N. J. L. Megson, Butterworth, London, 1958.
Laminated Plastics, by Duffin, Reinhold, New York, 1966.

CHAPTER 4

Amino Resins and Moulding Powders

UREA–FORMALDEHYDE (u.f.) resins began to assume commercial importance after the First World War. The pattern of their development is confused; they were at first seen by some workers as suitable materials for the production of an organic glass and many combinations of ingredients were tried, including the use of thiourea. The first moulding powder in this country was a mixed urea–thiourea–formaldehyde condensate and was marketed by British Cyanides Ltd. (now B.I.P. Chemicals Ltd.) in 1926. The I.G. Group in Germany produced urea based adhesives and resins for paint at about the same time, while the use of the resins to impart crease resistant properties to textiles was worked out by Tootal, Broadhurst and Lee. There was limited use of urea resins for production of decorative laminates in the mid-twenties but this outlet did not make much progress until later.

Melamine–formaldehyde resins were discovered in Germany in the early thirties but there was no commercial development in this country until after the Second World War.

Amino resins have a great advantage over phenolics in that they can be made colourless. By the use of suitable dyes and pigments, therefore, light-coloured articles and laminates can be fabricated.

RAW MATERIALS

Urea is a white crystalline solid, very soluble in water, and large quantities of it are used as a fertilizer throughout the world, especially in warm, dry climates. It is usually manufactured in the same location as ammonia by reacting some of the ammonia produced with carbon dioxide, which is available in large quantities as a by-product from the ammonia synthesis. The equation for the formation of urea may be written:

$$CO_2 + 2NH_3 \longrightarrow CO(NH_2)_2 + H_2O$$

The reaction takes place in the liquid phase at about 150°C and 150 atm pressure; ammonium carbamate, NH_2COONH_4, is an intermediate product. I.C.I. is the only U.K. manufacturer of urea, with a plant at Billingham, Co. Durham, of 350,000 t/a capacity, much of which is exported.

Melamine is a cyclic triazine derivative and, like urea, is a white crystalline solid, only very slightly soluble in water. Manufacturing processes for it are complex and no really cheap method of making it has yet been devised. The standard method is from calcium carbide via cyanamide and dicyandiamide according to the following series of reactions:

$$CaC_2 + N_2 \longrightarrow CaCN_2 + C$$
$$CaCN_2 + H_2O + CO_2 \longrightarrow N{\equiv}CNH_2 + CaCO_3$$

Cyanamide

$$2N{\equiv}CNH_2 \longrightarrow H_2NC = NC{\equiv}N$$
$$\qquad\qquad\qquad\qquad\quad |$$
$$\qquad\qquad\qquad\qquad NH_2 \qquad \text{Dicyandiamide}$$

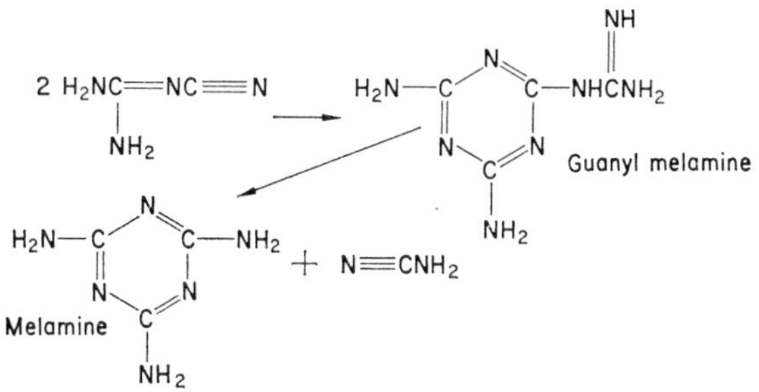

The conversion of dicyandiamide to melamine is carried out by heating with excess ammonia under pressure. An apparently cheaper method is the cyclization of three molecules of urea with the elimination of three molecules of water and this process is now coming into use on the large scale. The conversion is not so simple, however, as it appears on paper and is accompanied by much reconversion of urea to ammonia and carbon dioxide. There are two manufacturers of melamine in the United Kingdom, British Oxygen Chemicals Ltd., at Chester-le-Street, Co. Durham, with capacity of about 5000 t/a and Cyanamid of Great Britain Ltd. at Gosport, Hants., with about 3000 t/a.

CHEMISTRY OF AMINO RESINS

When urea and formaldehyde are mixed in aqueous solution under faintly alkaline conditions methylol urea is first formed.

$$NH_2CO\ NH_2 + HCHO \longrightarrow NH_2CONHCH_2OH$$

In the presence of sufficient formaldehyde the second amino group will react to produce dimethylol urea. If the condensation is carried out under acid conditions methylene urea, $NH_2CON=CH_2$, is formed; it is also produced when methylol urea is heated above its melting point.

$$\underset{\text{Methylol urea}}{NH_2CONHCH_2OH} \longrightarrow \underset{\text{Methylene urea}}{NH_2CO-N=CH_2} + H_2O$$

The methylol ureas may be separated as white crystalline solids which, on heating, evolve water and formaldehyde and change into soluble polymers. The reactions leading to formation of infusible, insoluble resins are not precisely understood and two possible mechanisms have been suggested.

The first postulates the production of chains of methylene urea molecules from the methylol ureas, the water and formaldehyde formed during these condensations keeping the system homogeneous. Cross-linking of chains then begins through further reaction with formaldehyde.

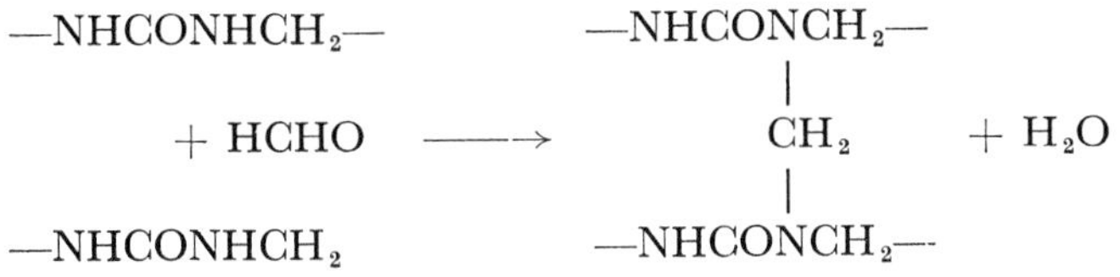

$$
\begin{array}{c}
-\text{NHCONHCH}_2- \\[4pt]
+ \text{HCHO} \\[4pt]
-\text{NHCONHCH}_2
\end{array}
\quad\longrightarrow\quad
\begin{array}{c}
-\text{NHCONCH}_2- \\
| \\
\text{CH}_2 \\
| \\
-\text{NHCONCH}_2-
\end{array}
\quad + \text{H}_2\text{O}
$$

and eventually leads to a cross-linked resin with, possibly, the following structure:

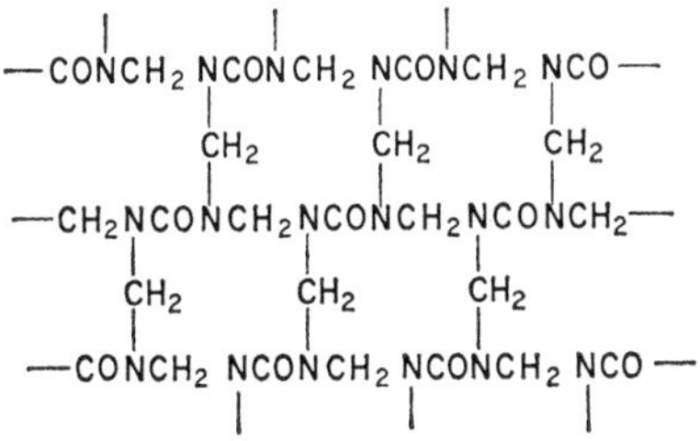

$$
\begin{array}{c}
| \qquad\quad | \qquad\quad | \\
-\text{CONCH}_2\ \text{NCONCH}_2\ \text{NCONCH}_2\ \text{NCO}- \\
\text{CH}_2 \qquad \text{CH}_2 \qquad \text{CH}_2 \\
-\text{CH}_2\text{NCONCH}_2\text{NCONCH}_2\text{NCONCH}_2- \\
\text{CH}_2 \qquad \text{CH}_2 \qquad \text{CH}_2 \\
-\text{CONCH}_2\ \text{NCONCH}_2\ \text{NCONCH}_2\ \text{NCO}- \\
| \qquad\quad | \qquad\quad |
\end{array}
$$

Formaldehyde can react with all three amine groups of melamine to form methylol compounds which then condense further in the manner suggested for methylol ureas to form cross-linked resins.

MANUFACTURING PROCESSES

Both urea and melamine resins are made in equipment similar to that described for phenolic resins on p. 68. A simple urea–formaldehyde resin may be obtained by boiling one molecular equivalent of urea with two molecular equivalents of formaldehyde in aqueous solution, with a trace of an acid catalyst such as formic acid. Formaldehyde containing the required amount of acid can be produced directly by oxidation of methanol. The solution thickens but remains homogeneous and miscible with water for some time, but water and formaldehyde gradually separate out and a hard, infusible resin is formed.

The condensation can be stopped at any time by adding alkali and, by choosing the right moment, a syrup can be obtained which is suitable for many applications. Spray drying of the syrup under vacuum gives a powdered resin which, provided the condensation has not been carried too far, will redissolve in water. Higher molar ratios of formaldehyde (above 2:1) increase the water solubility and the stability in solution but spray drying conditions become more critical and the solid cakes more readily. Urea resins have important applications in paints and lacquers, for which purpose the resin must be soluble in paint solvents and compatible with oil modified alkyd resins. This can be achieved by forming an ether with, for example, n-butyl alcohol. A suitable resin may be made by reacting urea, formaldehyde and n-butyl alcohol together under slightly acid conditions. After the reaction has proceeded to the desired stage, water is distilled off as its azeotrope with n-butyl alcohol and a water-free butylated resin is obtained in solution in n-butyl alcohol. This can be used in formulating paint resins. In practice, far more melamine resins than urea resins are used in this application.

Production of Moulding Powders

Resins for formulation into moulding powders are usually made with a urea–formaldehyde molar ratio of 1:1·5; contrary to the practice with phenolics, it is not necessary to produce a powdered resin in order to formulate the moulding powder. A neutral syrup containing 60–70% solid resin is filtered and then blended with fillers, such as bleached sulphite wood pulp, in a kneading machine, together with about 0·5% of a mould lubricant such as zinc stearate. For a good strength product the quantity of filler is about one-third to a half the resin content of the syrup; the lower proportions are used where especially easy flow properties are desired in the finished powder. Wood flour can be used for the cheaper products and regenerated cellulose is used for translucent products such as buttons. The kneaded product is oven dried under controlled conditions until it is just brittle and disintegrated; it is then blended with a catalyst, a stabilizer and any required pigments or dyestuffs in a ball mill and ground to a fine powder. Powders produced in this way have a low apparent density which may be increased by fluxing on mixing rolls (as is done with phenolics) and regrinding the plastic sheet so formed. An outline diagram of the process is shown in Fig. 18.

The properties of the catalyst are of great importance. It must be a compound which is inactive at ordinary temperatures but which will liberate sufficient acid at moulding temperatures to bring about the cross-linking of the resin; some examples are phthalic anhydride, and zinc sulphite and sulphate. The stabilizer is often hexamine, which neutralizes any acid liberated prematurely. It is essential that the catalyst chosen should not react adversely with the pigments or dyestuffs and that it should not corrode the mould.

U.F. RESINS AS ADHESIVES

Urea–formaldehyde resin based adhesives are used in large quantities for plywood production throughout the world. For this purpose a liquid adhesive is required and this is usually supplied as a two part system consisting of a water soluble resin and a hardener which are mixed by the user before application. The resin is made by condensing urea and formaldehyde to near the limit of water compatibility so that it cures quickly. The hardener must liberate sufficient acid when mixed with the resin solution to ensure that the resin is fully cross linked and must give a quick cure while having a reasonable pot life. The pot life is the time between the mixing of the ingredients and the point when the adhesive is just unusable and it must obviously be long enough to permit preparation and application of the adhesive in reasonable quantities. Ammonium chloride has been found to be a suitable hardener; it reacts slowly with excess formaldehyde in the resin, forming hexamine and liberating hydrochloric acid. Pot life can be increased, without a *pro rata* increase in curing

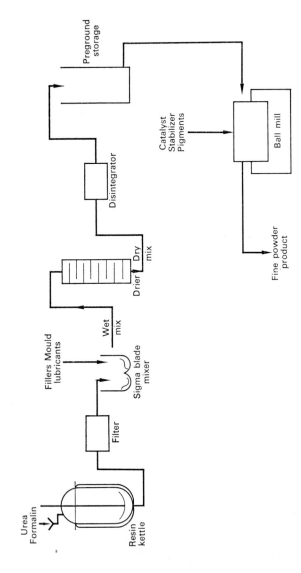

FIG. 18. Process diagram for preparation of u.f. moulding powder.

time, by the addition of hexamine. There is a tendency for the pH to decrease during cure due to reaction of liberated formaldehyde with more ammonium chloride, accompanied by production of further hydrogen chloride. This is counteracted by addition to the resin of a buffer compound such as sodium formate.

Adhesives formulated as described are excellent if the glue line is thin; where the surface is uneven, or a thick glue line is necessary for other reasons, the resin tends to craze. This may be prevented by the use of a gap-filling cement which is made by the addition of an extender and a filler. Extenders are also used in non-gap filling cements since they can cheapen the adhesive, improve spreading and prevent the resin adhesive penetrating porous surfaces; typical materials are bentonite and rye or wheat flour, used in quantities up to 20%. Fillers are normally cellulosic materials; in suitable cases their use actually improves the adhesion as well as filling gaps and preventing crazing. To be effective, however, the cellulose must be treated with an organic alcohol; benzyl or furfuryl alcohols are normally used.

Urea–formaldehyde glues are only suitable for bonding cellulosic materials such as wood, paper or cork; by appropriate formulation they may be cured hot or cold. As wood glues they are superior to phenolic resin based glues in that the water content of the wood is far less critical; they are completely resistant to fungi, woodworm and termites, although, if farinaceous extenders have been added, resistance to fungi may need to be strengthened by incorporation of fungicides. Simple u.f. glues have, however, only fair resistance to hot water and break down completely in boiling water; heat and water resistance may be improved by incorporating melamine or resorcinol–formaldehyde resins at, of course, increased cost. A glued joint made and tested according to B.S. 1204 gives a strength of 400–500 lb/in².

Foams

Urea–formaldehyde foams can be produced by vigorously agitating an aqueous resin with a foaming agent, such as naphthalene sulphonic acid, and an acid hardener, such as phosphoric acid, and discharging into a mould. The foam formed sets rapidly and is then allowed to dry for a few days at 40°C. The foams have densities between $0\cdot3$ and 1 lb/ft^3; they have not been developed as fully as polystyrene and polyurethane foams but show promise in building insulation. They can be made *in situ* and are less flammable than other foams.

Chipboard

Production of chipboard is one of the largest outlets for these resins; it has grown rapidly in recent years as shown in Table 5:

TABLE 5. U.K. OUTPUT OF CHIPBOARD
(000 long tons/annum)

1958	34·8	1963	91·2
1959	37·2	1964	132·6
1960	34·8	1965	147·6
1961	48·0	1966	162·0
1962	73·2	1967	147·6

Chipboard is made by disintegrating wood from forest thinnings and blending with a 60% u.f. syrup so that the product will contain 5–7% weight of dry resin. It is then either loaded into a hot mould and pressed into sheets of convenient size, or the damp mix is forced between parallel plates by a hammer action to give a continuous board of the desired thickness, which is then cut into suitable lengths; in either case the raw sheet is oven cured.

Application to Textiles

Very lightly condensed u.f. resins, which have not progressed far beyond the methylol urea stage, are used to give crease resistance to rayon and, sometimes, to cotton goods. 10–15% is often added to rayon but much less, around 5%, to cotton; addition of the resins to cotton causes a marked decrease in tensile strength and a 10%

addition can reduce the strength by half. The lightly condensed resins for rayon are often made *in situ* by the textile finishing companies. They are also produced for sale by the major resin manufacturers when some methylated spirit is added and they are concentrated under vacuum to give a resin with adequate shelf life. Cyclic derivatives of urea, e.g. ethylene urea, are often used in modern textile finishes to give crease resistance and resistance to bleaching agents based on chlorine.

Application to Paper

Urea–formaldehyde resins are used on a substantial scale to improve the wet strength of paper, mainly paper for packaging. To improve adhesion to the pulp and so prevent its being lost in the effluent water, resin with positively charged particles is used; these adhere to the negatively charged particles of pulp. The resin is cured as the paper passes over hot rollers during the drying process. The resins are prepared by modifying a u.f. resin with about 10% (calculated on the urea) of an organic base, such as tetraethylene pentamine, guanidine or triethanolamine and hydrochloric acid. An acid resin is produced, which is neutralized to give a good shelf life, but which is made slightly acid again before being used to treat the paper.

Limited use has been made of anionic resins made by condensing urea and formaldehyde in the presence of 10% (calculated on the urea) of sodium metabisulphite. Alum must be added to the paper suspension to precipitate the resin on the fibre. The method is very cheap and gives fair results on Kraft paper but not with other types of pulp.

MELAMINE–FORMALDEHYDE (M.F.) RESINS

There is a close resemblance between urea and melamine resins, both in the manner of their production and in their uses; this section will be devoted mainly to emphasizing their differences. All three NH_2 groups can react with two molecules of formaldehyde and melamine behaves as a hexafunctional compound. The melamine–formaldehyde condensation is very sensitive to pH; it is carried out at pH 7 when three molecules of formaldehyde are required to react with one of melamine and at pH 9–10 when higher proportions of formaldehyde are required; resins are most stable at about pH 10 and excess of formaldehyde increases the rate of resinifaction. Melamine–formaldehyde molar ratios of about 1:3 are generally used; the resins are more reactive than urea resins when heated and are less dependent on catalysts for curing.

Cellulose is the preferred filler for melamine moulding powders and the method of production is basically similar to that of urea products described on p. 83. Catalysts are not essential but small amounts of the same type as used for urea may be added to accelerate the cure. Melamine resins are less stable in solution than the corresponding urea compounds so, for sale, the syrups are spray dried or the solvent evaporated under vacuum. Laminating resins are then made by dissolving the dried powder in a mixture of water and alcohol to give a solution with 40–50% solids content; this has a short pot life and must be used quickly. Melamine laminates have very good heat resistance and may be produced in stable pale colours; the curing process is, however, slow and difficult to control and this, coupled with their cost, has limited their use to the surface layers of decorative board. Glass cloth laminates may be made using m.f. resins; these can be press cured at moderate pressures around 500–1000 lb/in^2 and the products have excellent heat resisting and electrical properties.

Etherification with n-butyl alcohol, as described for urea resins, gives products soluble in paint solvents and compatible with oil-modified alkyds which are valuable as lacquer constituents; the same process, using substantial quantities of methyl alcohol, increases the stability of melamine resin solutions for other purposes, such as formulation of adhesives. M.f. adhesives do not cold set and are only made on a modest scale; commercial urea adhesives are, however, often upgraded by the incorporation of a proportion of melamine as hardener.

Melamine resins are also used instead of urea resins for increasing the wet strength of high quality paper. In this case, also, positively charged resin particles are advantageous; they may be produced by dissolving a spray dried resin in water and adding dilute aqueous hydrochloric acid, when a colloidal condensation product with positively charged particles is formed.

Small quantities of melamine resins are used in textile applications. Although much superior to urea resins for this purpose, they are so much more expensive that they are mainly used for upgrading urea-based products. The resins also have a use as cross-linking agents for acrylic resins used in stiffening textiles, in surface coating applications and for coating paper and textiles.

The main advantages of melamine resins which justify their use, in spite of their high cost, may be summarized as follows: their heat resistance is better than that of phenolics while they are superior to urea resins in both water and heat resistance. This, combined with their pale colour, justifies their use in many decorative laminates and moulded products. Their electrical properties, too, are exceptionally good and, in particular, their resistance to surface tracking is outstanding among thermosetting resins.

PRODUCTION AND END USES OF AMINO RESINS

It is not possible to obtain figures for production of urea and melamine resins separately; the latter are, however, only a small proportion, probably 7–8%, of the total. U.K. production of all amino resins over the last 14 years is shown in Table 6.

The figures include the water or solvent content of liquid resins and it is worth noting that, in 1966, nearly 80% of production was as straight liquid resin with, perhaps, 50–60% dry resin content. Of this 80%, about 47% went into moulding powders and about 20% into chipboard, the remainder being used in textile and paper treatment. Of the remaining 20% of total production, about half were butylated resins in liquid form for the paint industry and the remainder solid resins, mainly for adhesives and laminates.

TABLE 6. TOTAL PRODUCTION OF
AMINO RESINS IN THE U.K.
(000 long tons)

1954	57·2	1961	80·0
1955	62·4	1962	91·0
1956	64·4	1963	96·8
1957	67·0	1964	112·9
1958	65·1	1965	115·7
1959	71·7	1966	119·1
1960	80·1	1967	116·6

In addition to B.I.P. Chemicals Ltd., who, as noted on p. 84 were the first producers of amino resins in this country, the principal U.K. producers are C.I.B.A. (A.R.L.) Ltd. manufacturing amino resins, especially melamine resins, for their own use. As noted on p. 93 under "Textile Applications", small users may produce their own resins on the spot as an alternative to buying ready made resins.

READING LIST

Aminoplastics, by C. P. Vale and W. G. K. Taylor, Iliffe Press for the Plastics Institute, London, 1964.
Synthetic Resins and Allied Plastics, by R. S. Morrell, Oxford University Press, London, 1951.

CHAPTER 5

Unsaturated Polyester Resins

DEVELOPMENT of the modern unsaturated polyester resin is generally attributed to Carleton Ellis, who had patented the production of polyester resins for use as lacquers as early as 1922, although the patent was not published until 1933. The idea of cross-linking a maleic polyester by copolymerization with styrene was also due to Carleton Ellis, the patent being published in 1940. Commercial development started in the United States during the Second World War, and in this country soon after the war ended. The early resins used allyl esters, such as diallyl phthalate or allyl diglycol carbonate, as the cross-linking monomers and the former is still the preferred compound for the so-called alkyd moulding powders.

Unsaturated polyesters have come to be associated with glass fibres in the production of reinforced plastics. The chemical inertness, great strength and good temperature resistance of glass fibres have always made them attractive as a reinforcing material for laminated plastics in place of the cloth and paper commonly used. Laminating resins based on phenols, urea or melamine, however, must be cured at substantial pressures, often exceeding 1000 lb/in^2, to prevent blistering by vaporization of the water of condensation formed during the cure. Glass fibre is damaged by these high pressures and could not, therefore, be used. The outstanding advantage of the unsaturated polyester is that the cross-linking reaction is essentially an addition polymerization (p. 22), which takes place without elimination of water or other small molecule. This means that only sufficient pressure to keep the object being made to the desired shape is required during

97

the curing operation and expensive moulds are no longer necessary. This opens the way to the building-up of large objects from
reinforced plastics and enables the potentialities of glass fibre as a
reinforcing material to be realized. Methods for the commercial
production of glass fibre in a form suitable for the manufacture of
high-strength reinforced plastics were developed at about the
same time as the unsaturated polyester resins; the development
appears to have been stimulated by the increasing demand and
might well have been earlier or later if the timing of the demand
had been different but, because they were developed at the same
time, glass fibre and polyester resins have come to be looked on as
natural associates. The growth of output over the last 14 years is
shown in Table 7 below; the average growth over this period has
been nearly 40% per annum and, although this rate will clearly
decrease, there appears still to be plenty of room for further
expansion.

TABLE 7. U.K. PRODUCTION OF
UNSATURATED POLYESTER RESINS
(figures in 000 t/a)

1954	0·5	1961	9·8
1955	1·0	1962	11·5
1956	2·2	1963	13·7
1957	3·4	1964	18·4
1958	4·7	1965	24·2
1959	6·6	1966	25·5
1960	8·8	1967	27·3

CHEMISTRY OF THE POLYESTER RESINS

The essential constituents of a thermosetting polyester system
are a dihydroxy alcohol and a dicarboxylic acid, one of which
must be olefinic—i.e. tetrafunctional—in order to provide cross-
linking points in the polymer molecule; mixed acids or diols may
be used of which only a proportion is olefinic. In addition a
polymerizable olefin must be added, together with a suitable
catalyst, to act as a cross-linking compound.

A detailed description of the method of formation and cross-linking of an unsaturated polyester has been given on p. 21 to illustrate the effect of functionality. Although the methods of formation and cross-linking of all unsaturated polyesters are basically similar, the properties of the resin and, equally important, its price are greatly affected by the combination of reactants used. A commercial resin formulation is almost always a compromise which seeks to obtain the optimum combination of resin properties at a price which the market will bear.

If a polyhydric alcohol is used as one of the reactants the hydroxyl groups may take part in the formation of cross-links between the chains. A straight chain polyester made from phthalic anhydride (P) and a dihydric alcohol (A) may be represented:

$$—A—P—A—P—A—P—A—P—A—P—A—$$

If, however, A is a trihydric alcohol the product may be:

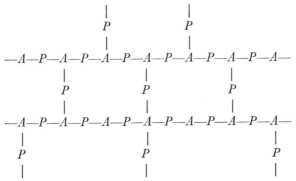

This is the basic reaction for the production of the alkyd resins, also known as glyptals, for which the reactants are usually phthalic anhydride and glycerol; but other polyhydric alcohols, for example pentaerythritol, may be used. Oil modified alkyds are produced by adding to the reaction mixture drying oils, consisting essentially of the glyceryl esters of unsaturated long chain fatty acids. The products are usually viscous liquids which may be compounded with pigments, solvents and the usual cobalt, lead and manganese based paint driers to form paints. For air drying paints the unsaturated fatty acid component must contain two olefinic groups while stoving finishes require only one. United Kingdom production of alkyds in 1967 amounted to some 64,000 tons, nearly all for production of paints, which are outside the scope of this book. Some solid alkyds are, however, used on a small scale for moulding and are described on p. 119.

RESIN PROPERTIES AND FORMULATION

There are no readily available unsaturated dihydric alcohols and almost all commercial resins are based on unsaturated dicarboxylic acids, mainly maleic, in the form of its anhydride; its geometric isomer fumaric acid, which does not form an anhydride, is sometimes used. A number of saturated dihydric alcohols are commercially available, of which ethylene and propylene glycols are produced in the largest quantities. Polyhydric alcohols can also be used; good examples are pentaerythritol, sorbitol and glycerol, although the last named is chiefly used in alkyds for the paint industry.

Maleic anhydride and ethylene glycol form the simplest unsaturated polyester:

$$- CH{=}CHCO{-}O{-}CH_2CH_2{-}O{-}OCCH{=}$$
$$CHCO{-}O{-}CH_2CH_2{-}O{-}$$

This polymer is, however, crystalline, i.e. the polymer chains tend to arrange themselves in a regular formation; it is hard and tough, shows high shrinkage on cooling and is virtually incompatible with olefinic cross-linking agents such as styrene. Crystallinity can be reduced and compatibility with styrene improved by replacing part of the maleic anhydride by a larger molecule of a saturated anhydride, such as phthalic anhydride. Equimolar proportions give good results; too much phthalic leads to a resin that has too low a viscosity and does not cure fully, so that the cured resin has a low softening point. Crystallinity may also be reduced by replacing some, or all, of the ethylene glycol by propylene glycol; complete replacement, however, reduces the compressive strength of the cured resin. It is usual for the glycols to be present in about 10% molar excess over the combined acidic components since some of the glycol component is lost by evaporation during esterification.

Styrene is the most favoured olefin for cross-linking and is used in the ratio of about one mole of styrene to two moles of the combined acids. Higher styrene contents give very low viscosity resins, with the cured product showing high shrinkage on cooling, while too little styrene gives resins which do not develop fully thermosetting properties on curing.

Other raw materials are used to confer special properties on the finished product. Thus, di- or tri-ethylene glycols in place of the monoglycol increase the flexibility of the cured resin while partial replacement of a glycol by diphenylol propane improves the chemical resistance. The strength of the resin in a glass-reinforced product is of secondary importance since great strength is given by the glass fibres; in the less well developed casting applications, however, the resin must have good mechanical properties. With maleic–phthalic anhydrides as the acid components, propylene glycol gives the maximum flexural strength, tri-methylene glycol the highest compressive strength and 1,5 pentane diol the best impact strength.

In reinforced plastic products the water absorption and heat distortion temperature are especially important characteristics. Ethylene glycol itself gives products with very low water absorption and propylene glycol is almost as good. Longer chain glycols, or compounds with ether linkages such as di- and tri-glycols, are considerably inferior. Heat distortion temperatures are markedly affected by the degree of unsaturation of the acid components and increase with increasing maleic content. Propylene glycol with a 1:1 molar ratio of maleic–phthalic anhydrides gives a product with a heat distortion temperature of about 70°C but, by raising the maleic content of the anhydride mixture, figures of over 100°C may be obtained while still higher figures may be reached if part of the propylene glycol is replaced with diphenylol propane.

Resin properties may also be modified by changes in the acid components. While the maleic–phthalic combination is by far the most commonly used, the phthalic anhydride can be replaced by a long chain dicarboxylic acid, such as adipic or sebacic acid, to give flexible resins. Isophthalic acid too, may be used; it gives a resin which is marginally tougher and more heat resistant than the phthalic anhydride based resins, as well as having lower water absorption. Replacement of maleic anhydride by fumaric acid gives a tougher and more heat resistant, but partly crystalline, resin. There is evidence from infrared spectra that, in practice, the maleic does isomerize to the *trans*-form during the resin forming process; the resins made from the two acids do, however, differ significantly in their properties, possibly due to differences in the arrangement of the unsaturated acid molecules along the chain.

While the cross-linking agent is usually styrene, other olefinic compounds may have advantages. Thus, in the production of reinforced sheets for roof lights, the glass fibre can be made almost invisible by using a 4:1 mixture of styrene and methyl methacrylate; this gives a resin of refractive index almost identical with that of glass. Methyl methacrylate, however, has the disadvantage of slowing down the curing rate. Diallyl phthalate, because of its tetrafunctionality, gives a very highly cross-linked resin and a tough product but it is much more expensive than styrene. Triallyl cyanurate gives a resin of maximum temperature resistance. Vinyl toluene has been used to some extent as a replacement for styrene; it has the advantage of being less volatile but it does not cross-link so rapidly.

A list of available diols, di-acids and cross linking agents follows:

Diols

CH₂OH
|
CH₂OH Ethylene glyco

CH₂OH
|
CH₂ Diethylene glycol
|
O
|
CH₂
|
CH₂OH

CH₂OH
|
CH₂ Trimethylene glycol
|
CH₂OH.

CH₂OH
|
CHOH Propylene glycol
|
CH₃

CH₂OH
|
CH₂
|
CH₂ 1,5 Pentane diol
|
CH₂
|
CH₂OH

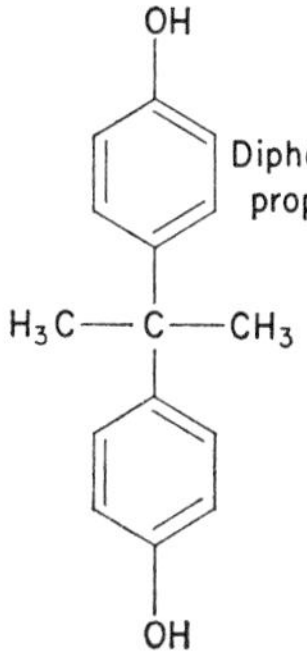

Diphenylol propane

H₃C—C—CH₃

Di-acids

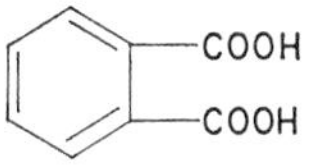

o-Phthalic acid

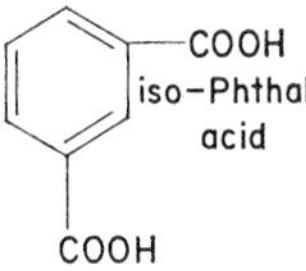

iso-Phthalic acid

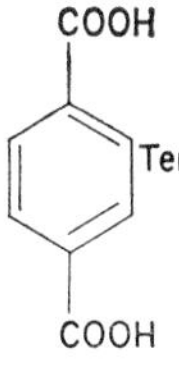

Terephthalic acid

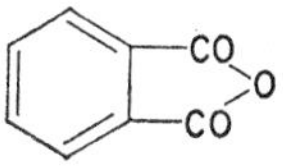

Phthalic anhydride

CHCOOH Maleic acid
‖
CHCOOH

HOOCH Fumaric acid
‖
HCCOOH

HOO(CH₂)₄ COOCH
Adipic acid

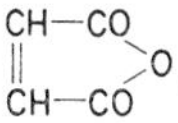

Maleic anhydride

HOOC(CH₂)₈ COOH
Sebacic acid

Cross-linking agents

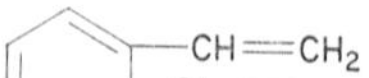

$-CH{=}CH_2$
Styrene

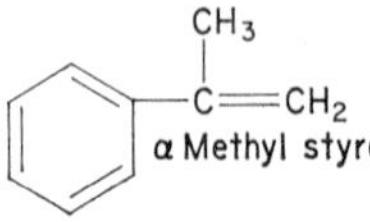

α Methyl styrene

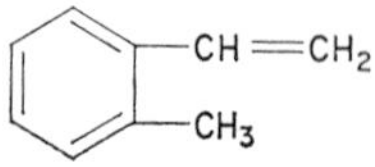

$-CH{=}CH_2$
$-CH_3$ o-Vinyl toluene

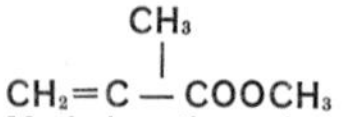

CH_3
$CH_2{=}C{-}COOCH_3$
Methyl methacrylate

$-COOCH_2\ CH{=}CH_2$
$-COOCH_2\ CH{=}CH_2$

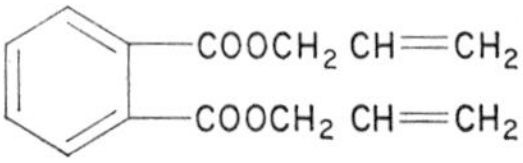

Diallyl phthalate

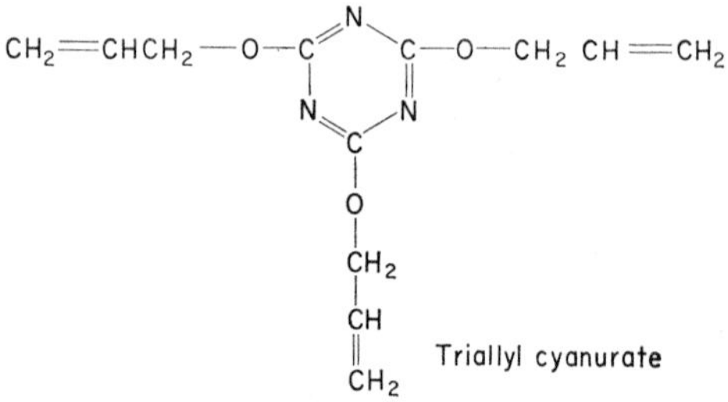

Triallyl cyanurate

Fire Resistance

Products made from ordinary unsaturated polyesters reinforced with glass fibre are not highly flammable but, although the cured resin will not melt, it is decomposed by excessive heat and will burn slowly. Use of these materials in the building industry and for many domestic articles has led to efforts to make them more flame resistant. Antimony oxide mixed into the resin, generally in association with chlorinated wax, greatly slows down the rate at which the resin burns; the finished glass reinforced product is, however, opaque and the chlorinated wax tends to sweat out of the laminate. These admixtures always reduce the flexural strength, sometimes by up to 50%. Introducing chlorine into either the saturated or the unsaturated acid component will give substantial fire resistance. Thus, tetrachlorophthalic anhydride can be used but chlorinated wax, or some similar material, is also needed with the disadvantages already mentioned.

Carrying out a Diels–Alder reaction with the maleic anhydride and hexachlorocyclopentadiene gives chlorendic anhydride, also known as HET acid.

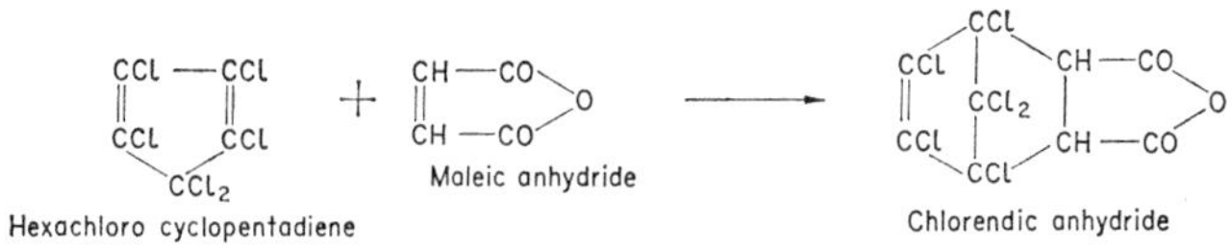

In spite of its double bond, HET acid reacts virtually as a saturated acid and must be considered as a replacement for the phthalic anhydride. This acid gives an exceptionally fire resistant resin which is, however, brittle and yellows rapidly. In practice, therefore, such resins have only part of the phthalic anhydride replaced by HET acid and have a compound which absorbs ultraviolet light, such as a substituted benzoquinone, incorporated in them. Development work is taking place with fluorinated acids, instead of the chlorinated compounds used until now, and with phosphorus compounds such as diallyl-benzenephosphonate as cross-linking agents. All of these expedients greatly increase the cost of the finished product so that they tend to be used only where cost is not a major factor.

Pre-impregnation of Glass Cloth

One of the disadvantages of the earlier polyesters was the necessity of laying up the glass cloth or mat in the desired form, impregnating it with the resin–cross-linking agent–catalyst mixture and curing in a closely integrated operation. It was not possible to prepare impregnated sheets which could be stored dry until required as can be done with phenol and urea based laminating resins. Two techniques have recently been developed to overcome this disadvantage. In the first a high viscosity resin is prepared by replacing some of the glycol in the normal formulation by pentaerythritol; this resin is then dissolved in acetone containing diallyl phthalate and benzoyl peroxide. This solution is then used to impregnate the reinforcing material, the acetone is evaporated at a low temperature and the dried sheet, known colloquially as a "prepreg", stored until required. An alternative method, when a filler is of no disadvantage, is to blend the catalyst with a filler before stirring both into a conventional resin. Material impregnated with this mixture can then be stored, as the catalyst is not effective until heat is applied.

Resistance to Sunlight

Resins to be used for making roof sheeting material must not discolour in strong sunlight. Such resins are normally stabilized by addition of $0 \cdot 1 - 0 \cdot 5\%$ of an ultraviolet light absorbing compound such as a benzophenone; certain salicylates are also used.

Curing

Curing is the process in which the polyester chains are linked together at their points of unsaturation by a compound with an olefinic group. The reaction is essentially an addition polymerization promoted by a free radical mechanism as described in Chapter 1. From the molar proportions given on p. 101 it can be seen that, on average, there will only be one olefinic molecule forming the bridge between each pair of unsaturated points in the polyester chains.

According to the application the resin may be required to cure hot or cold; even when hot curing is required, it is usual to add a catalyst to speed up the cross-linking reaction. Cold curing resins are usually catalysed by an organic hydroperoxide activated with a cobalt salt or a peroxide associated with a tertiary amine. Cyclohexanone or methyl ethyl ketone hydroperoxides with cobalt naphthenate are typical of the first system and benzoyl peroxide with dimethyl aniline of the second. The cobalt salt and the tertiary amine may both be used in certain cases. Generally, cobalt cured systems cure completely through the mass of resin, although the process may take several weeks before it is finally completed, while amine assisted systems tend to give products which craze with age, mainly because they are under-cured. The attraction of the amine assisted system is that either the peroxide or the tertiary amine may be incorporated in the resin without initiating the cure.

For hot curing the catalyst is frequently benzoyl peroxide made into a paste with dibutyl phthalate. About 2% of the catalyst is usually added and curing temperatures of about 100°C are normal. The resin–catalyst mixture has a shelf life of 10 days or so; replacement of the benzoyl peroxide by ditertiarybutyl peroxide gives an even longer shelf life but a temperature of 150°C is required for curing.

A practical problem in the production of resin formulations is that mixing processes, such as incorporation of the cross-linking agent, may have to be carried out at elevated temperatures because of the high viscosity of the resin and there will be a tendency for curing to start during the mixing process. The effect may be prevented by addition of a small quantity of an inhibitor such as 0·01–0·05% of hydroquinone or tertiary butyl catechol, which also improves shelf life.

The curing reactions of unsaturated polyesters are exothermic so that the temperature of the resin mass rises during cure. If this temperature rise is too great the properties of the cured resin will be adversely affected and this presents practical problems of heat dissipation where bulky objects are being made. The temperature rise can be substantially reduced without significantly prolonging the curing time, by replacing part of the styrene normally used by alpha methyl styrene; this also reduces shrinkage during cure.

Air or oxygen have an important effect on the curing of unsaturated polyesters; the resins are always manufactured in an inert atmosphere of nitrogen or carbon dioxide as they discolour badly if air is present. In spite of the use of peroxides to catalyse the curing process, contact with the air actually slows the cure; the body of the resin may cure satisfactorily but the surface tends to remain tacky. This is believed to be partly due to evaporation of the styrene, or other cross-linking agent, thus causing the surface layer to become deficient in this material so that it does not cure properly. This may be overcome by incorporating a small amount of paraffin wax into the resin formulation; the wax is miscible with the uncured resin but, as the cure proceeds, the wax separates out on the surface, forming a protective layer against the air and preventing evaporation of volatile constituents of the resin mix. Evaporation is also minimized by the use of high boiling cross-linking agents such as diallyl phthalate or vinyl toluene.

Methods of Production

Unsaturated polyester formation is a condensation reaction and the manufacturing processes and plant are similar to those already described for condensation polymers. A batch size of about 5 tons is usual and a catalyst, such as *p*-toluene sulphonic acid, may be used for the esterification reaction; water of condensation is distilled off, often as an azeotropic mixture with a solvent added for the purpose such as xylene. As already noted, arrangements are made so that a stream of nitrogen or carbon dioxide can be passed through the kettle during the process. The viscosity of the resin will depend on the chain length of the polyester and may be controlled by adding a chain terminating agent, such as a small quantity of a monofunctional acid or alcohol, at the required stage in the reaction. Incorporation of the cross-linking agent and other additives may be carried out in the resin kettle or in conventional mixing equipment.

Unsaturated polyester resins are normally viscous liquids; methods of testing resins are detailed more fully in Chapter 2. The following properties are important: colour, both for the sake of appearance and because a dark colour may indicate partial oxidation or contamination which will affect the setting time;

acid number as a measure of the completeness of esterification; viscosity, since it determines the ease with which the resin will penetrate the reinforcing material, and setting time, a property of great practical importance in production processes. Storage life and temperature rise during cure may be controlled by proper formulation but will obviously affect the organization of operations.

APPLICATION OF UNSATURATED POLYESTERS

There are three main areas of application for unsaturated polyester resins: surface coatings, potting and casting and in the production of glass-reinforced products. The first two have only developed to a very limited extent in the United Kingdom and will be briefly described first; the production of glass-reinforced products, which is by far the largest outlet for the resins, will then be dealt with in some detail.

Use for Surface Coatings

Unsaturated polyesters may be used to give very high-gloss surfaces on wood but have not made much progress in the United Kingdom. Their great disadvantage is that the gelling of normally formulated resins is inhibited by air so that films remain tacky on the surface for a long time. The incorporation of paraffin wax to prevent this, as described on p. 108, gives a dull surface which must be buffed, with a consequent increase in cost, in order to gain the advantage of the high gloss imparted by the polyester. Various modifications to the resin system have met with some success. Styrene may be replaced as cross-linking agent by a glycerol mono or diallyl ether, or its adipic acid esters, which are less easily lost from the surface by evaporation. Partial replacement of the phthalic anhydride by tetrahydrophthalic anhydride, while retaining the conventional cross-linking agents such as styrene, has given promising results. Most users have found, however, that even with these improvements the surface still has to be buffed, so the paraffin wax technique continues to be the most popular.

Potting and Casting

Potting is the encapsulation of delicate electrical or electronic components in a block of resin, thus providing both mechanical protection and insulation. The excellent electrical properties and low power factor of unsaturated polyesters make them particularly suitable for these applications and cold setting resin formulations have found uses in the electronics industry. More recently, however, they have tended to be displaced by epoxy resins for these purposes. Potting is, of course, merely a special case of the general technique of casting in which liquid resin is run into a mould of the desired shape and allowed to set. Another casting application is the manufacture of "pearl" buttons.

Glass Reinforced Products

The use of glass reinforced polyester resins is really a special application of the general technique of production of plastic laminates described in Chapter 3. At first sight there appears to be no reason why glass should be the only reinforcing material used but there are combined technical and economic reasons for this. The special attractions of glass fibre are its great strength combined with outstanding resistance to heat and corrosion. The only other reinforcing material which approaches glass in this combination of properties is asbestos and this has been used to some extent; it does not, however, have the strength of glass. Other high quality reinforcing materials, such as nylon and terylene, are as expensive as glass and do not have its unique combination of properties. On the other hand, glass, as already noted, cannot be used with the cheaper phenolic and urea laminating resins because it will not stand the pressures required. Thus, apart from the fact that their commercial development took place at the same time, the two materials make an ideal combination for the production of high quality reinforced plastic products.

The Glass Fibre

This is the more expensive component in the combination; it can be produced in forms having tensile strengths up to 400,000 lb/in^2 and has excellent dimensional stability, negligible water absorption, good electrical properties, is immune to biological attack and has good weathering resistance and durability. Its flex life is poor so it is not suitable for making products subject to continuous movement.

For reinforced plastics glass is produced in two forms: rovings and woven products. Rovings are formed when several strands of glass fibre are wound together without deliberately twisting so that the fibres lie more or less parallel. The rovings themselves can be woven to form a relatively low cost product with properties not greatly inferior to conventionally woven products and are also used as the reinforcement in the production of tubes and rods. By far their largest use, however, is for the manufacture of chopped strand mat. This is a mat composed of chopped strands of glass fibre about 2 in. long bonded together in a random manner. The bonding agent must be one which will not affect the subsequent processing.

Woven glass cloths are made by conventional methods in a number of different weaves. Plain weave is often used but higher strength products are produced by using a twill or a satin weave. Undirectional weaves can be obtained by using very little weft; this gives a cloth with the maximum strength in one direction, which is useful in some applications.

Ordinary textile sizes are often applied during production of the glass to reduce damage by abrasion. These must be removed by water washing or the use of solvents before the glass can be used in reinforced plastics. After desizing, special finishes may be applied to the glass to improve its bonding with the resin; typical ones are vinyl trichlorosilane or a chrome complex such as methylacrylato-chromic chloride. The former is believed to hydrolyse on to the glass surface leaving a vinyl group free to attach itself to the unsaturated polyester.

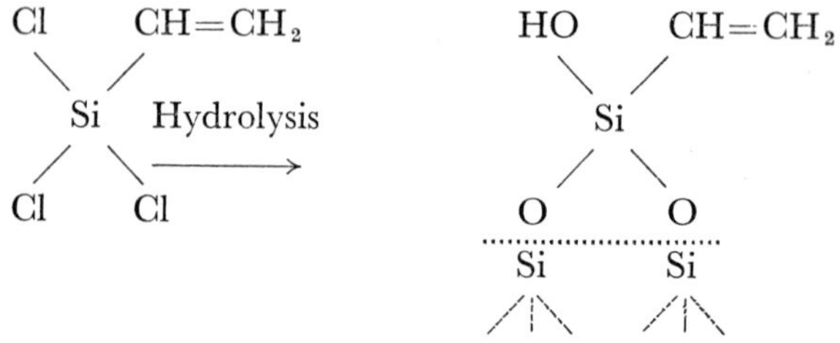

$$
\begin{array}{ccc}
\text{Cl} \quad \text{CH}=\text{CH}_2 & & \text{HO} \quad \text{CH}=\text{CH}_2 \\
\diagdown \quad \diagup & & \diagdown \quad \diagup \\
\text{Si} \quad \text{Hydrolysis} & & \text{Si} \\
\diagup \quad \diagdown \longrightarrow & & \diagup \quad \diagdown \\
\text{Cl} \quad \text{Cl} & & \text{O} \qquad \text{O} \\
& & \text{Si} \qquad \text{Si}
\end{array}
$$

The chromium complex attaches itself by a somewhat similar mechanism leaving the methylacrylato group to react with the resin.

Wet Lay-up Techniques

Laying-up is the process of supporting the glass reinforcing material in the required shape and impregnating it with the resin. Wet lay-up is the longest established technique and can be carried out with simple and inexpensive equipment; it has thus enabled reinforced articles to be made in a wide variety of shapes and of a size which would not have been possible with a phenolic or urea resin which required high pressure and an expensive mould. There are several variants of the wet lay-up method which is illustrated in Plates III–VII.

Single Mould Process

In this method a single male or female mould is used according to which side of the finished article is required to be smooth. The mould can be of any rigid material which will not be attacked by the constituents of the resin and need only be strong enough to carry the weight of the article while it sets; wood, plaster, cured glass reinforced polyester and some metals are all used according to circumstances. If the mould is made of a porous material it

PLATE III. Application of release
agent to mould surface. (By courtesy
of Scott Bader & Co. Ltd.)

PLATE V. Impregnation of glass mat
with resin. (By courtesy of Scott
Bader & Co. Ltd.)

PLATE IV. Application of gel coat.
(By courtesy of Scott Bader & Co.
Ltd.)

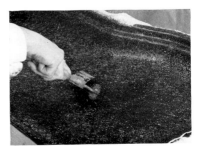

PLATE VI. Rolling of impregnated
mat. (By courtesy of Scott Bader &
Co. Ltd.)

PLATE VII. Final moulding and
mould. (By courtesy of Scott Bader &
Co. Ltd.)

must first be sealed and mould surfaces are treated with a mould release agent such as aqueous polyvinyl alcohol solution.

A coat of freshly mixed resin is then applied—frequently it is simply brushed on—and allowed to gel. This makes certain that the glass mat nearest to the mould surface is properly filled, otherwise the slight shrinkage of the resin which takes place on setting leaves the texture of the glass mat as a raised pattern. Glass mat of the required thickness is next laid on the prepared mould, and freshly mixed resin containing curing agent and accelerator applied by brush. The resin is worked into the glass fibre by hand to reduce to a minimum entrapped air cavities and the laminate is further consolidated by working over the surface with a roller. About 70 parts of resin by weight should be applied to every 30 parts of glass fibre.

A properly formulated resin will gel in 30–45 min and after 2–4 hr the article can be removed from the mould and set aside to "mature" while the cure is completed. This may take some days, although the time may be reduced to a few hours if the article can be heated to, say, 80°C. It will be clear that, for large articles, this method involves a carefully co-ordinated and timed series of operations with the setting time of the resin suitably adjusted to fit.

Pressure Bag Process

This is a slightly more complicated technique than the single mould process and is used when a greater rate of output is required and, usually, when hot curing can be applied. A bag of flexible material is placed against the free surface of the resinated mat and the bag blown up to force the impregnated mat against the mould surface. Alternatively a single sheet of flexible material is used and the space between it and the mould evacuated so that the article is forced into its mould by air pressure. The flexible material is normally heat resisting rubber, with a sheet of cellophane placed between it and the resinated glass, as rubber may be attacked by the resin.

The method of laying-up is otherwise similar to the single mould method. When the pressure has been applied, the whole assembly is placed in an oven or an autoclave for curing. The formulation of the resin is, of course, suitably adjusted for the conditions. An incidental advantage of this method is that both surfaces of the product are smooth.

Double Mould Method

In this technique both male and female moulds are used and give the advantage of smooth surfaces on both sides of the product. The glass mat is laid out on a sheet of cellophane and treated with the resin; a second sheet of cellophane is laid over it and the assembly is transferred to the mould. The other half of the mould is then placed in position and the two halves clamped together. The resin is allowed to set in the cold, the final stages being the same as for the single mould method. Quite complicated shapes can be made in this way; the products are of high quality but the rate of output is only of the same order as that obtained with the single mould method. An advanced application of this technique is illustrated in Plate VIII.

Products

A wide range of relatively large objects can be made by one or other of the three methods described above. These include such things as boat hulls, lorry cabs and car bodies. The physical properties of the product will depend greatly on the efficiency with which the processing, especially the impregnation step, has been carried out. Representative figures are given in Table 8.

Manufacture of Rods and Tubes

Rods are usually produced by passing a roving through a bath of resin and then forcing it mechanically through a heated die when the resin gels. The rod so formed is passed through a long

PLATE VIII. A double mould open showing the moulding, complete with bonded-in threaded inserts, being removed from the lower tool. (By courtesy of B.X.L. Plastics Materials Group Ltd.)

TABLE 8. PROPERTIES OF GLASS REINFORCED
MATERIAL FROM WET LAY-UP METHODS

Tensile strength	15,000 lb/in²
Compressive strength	20,000 lb/in²
Flexural strength	20,000 lb/in²
Flexural modulus	$1 \cdot 5 \times 10^6$ lb/in²
Impact strength (unnotched)	20 ft lb/in²
Density	$1 \cdot 6$

maturing oven to complete the cure. Rods with a strength of over 100,000 lb/in² can be produced and find a wide variety of uses, for example in the construction of high quality fishing rods.

Tubes are normally made by similarly impregnating the roving and then winding it under slight tension on a mandrel previously treated with polyvinyl alcohol or a similar release agent. The winding is carried out mechanically at a predetermined helix angle; when the desired thickness has been attained the tube is wrapped with cellophane and either allowed to cure cold with the mandrel still rotating, followed by a stoving operation, or the whole assembly is placed in an oven to cure. The resin mix must be correctly formulated according to the method adopted; resin drainage can be a problem when hot curing is employed. When completely cured the tube is stripped of its cellophane covering and removed from the mandrel.

Typical uses for glass reinforced polyester tubes are for chemical plant, for aerial and lifeboat masts and for piping of all kinds, especially in ships, where their non-magnetic properties are useful. The very high strength–weight ratio of reinforced polyester rods and tubes has encouraged their use in sports equipment. In addition to the fishing rods already mentioned, they are used for javelins and for vaulting poles; in the latter application their elasticity and lightness has added many inches to the maximum height which a particular pole-vaulter can attain. The tubes have high temperature resistance and will stand temperatures of at least 150°C, higher than conventional thermoplastics. An interesting new use is the cladding of concrete pipes to increase their bursting strength; in this application the concrete pipe itself forms the mandrel.

Mechanization of Wet Lay-up Techniques

The "bucket and brush" methods so far described are still very widely used, but they have the disadvantage of high labour content. For some applications, especially repetitive production of a single product, various mechanical expedients have been adopted to save labour. The most common is a spray technique in which glass fibre, resin, curing agent and accelerator are mixed in a spray gun and the mould covered to the required depth by spraying. The products are no better than those made by the simpler method, the sole advantage being in labour productivity; great skill is required in the spraying operation and the spray gun itself is of a very special design. Some advantage can be gained by laying-up the glass mat as previously described and then spraying the resin on to it; in any spraying technique, however, there is likely to be loss of cross-linking agent by evaporation during spraying and special resin formulation is required to provide for this.

Where large areas of a simple form, for example corrugated roofing sheets, are to be made, a continuous process can be operated. The glass mat, or chopped glass roving, is spread out on a moving belt of cellophane and impregnated from a traversing spray gun with a resin formulated to cure at elevated temperature. The impregnated mass then meets a second sheet of cellophane coming down from above and the uncured sheet, protected on both sides by cellophane, passes between rollers, or some similar device, to eliminate air bubbles and excess resin; the sheet then passes between corrugated formers and the whole assembly travels through a long oven to cure the resin. The cured sheet emerging from the oven can then be cut into convenient lengths.

By omitting the corrugated formers from the system the process may be adapted to produce flat sheet. A modification of it is being applied commercially to the production of a decorative laminate from ordinary cellulose paper and an unsaturated polyester. In spite of the relatively high cost of the polyester, the use of a continuous process and the avoidance of expensive pressing operations are believed to make the process com·· petitive with those described in Chapter 3.

Preform Moulding

This technique is used where articles of moderate size, such as safety helmets or washing machine bowls, are to be made. Glass roving is chopped into 2 in. lengths and carried by a current of air on to a porous screen shaped to match the final moulding. The screen can conveniently be made from mild steel plate, drilled or punched to leave 40% of its area open, and is coated with a layer of wax or polytetrafluorethylene to avoid sticking. A binder is applied to the glass mat thus formed as the glass does not "felt" when formed in this way; the binder may be starch, dextrin, polyvinyl acetate or other resins and the quantity does not normally exceed 5% by weight on the glass. The preform, still on its screen, is placed in an oven to dry and cure the binder, after which it can be handled. It is then placed in a heated metal mould, a predetermined charge of polyester resin is poured over it, the mould is closed and a moderate pressure of 100–200 lb/in^2 applied; this forces the resin into all the interstices of the glass preform and cures it. A mould lubricant is usually applied to the heated mould before the preform is placed in position.

Considerable variations of resin–glass ratio can be achieved by this technique; the usual wet lay-up ratio of 70% resin to 30% glass can be reversed, resulting in much stronger products.

Use of Prepregs

The preparation of prepregs has been described on p. 105. Glass cloth is usually used for prepregs as the dried impregnated cloth can be conveniently stored in a roll for up to 6 months, pieces being cut off as required. The cloth can be tailored to the shape required and as many layers as are necessary to give the desired thickness can be laminated together; curing is carried out in a double mould at a moderate pressure of about 100 lb/in² and a temperature of 100°C. The resin–glass cloth ratio is usually fairly low and very high strength products can be obtained by this method. Typical figures are given in Table 9.

TABLE 9. PROPERTIES OF CURED GLASS/POLYESTER PREPREGS

Tensile strength	40,000 lb/in²
Flexural strength	40,000 lb/in²
Impact strength (unnotched)	20,000 lb/in²
Dielectric strength	150 V/mil
Insulation resistance (after	
$2\frac{1}{2}$ hr in water)	5–20,000 megohms
Power factor (10^6 cycles)	0·03
Dielectric constant	4·0

Products made in this way are relatively expensive and find their main outlets in the electrical industries.

MOULDING APPLICATIONS

Only a very small proportion of the total output of unsaturated polyester resins goes into conventional moulding applications. The fact that they will cure at low pressures without elimination of water, however, makes them attractive for some purposes. Ordinary unsaturated polyesters may be converted to dough moulding compounds and some specially formulated alkyd resins also have small scale outlets in moulding powders.

Dough Moulding Compounds

A polyester resin chosen for manufacture of these compounds should have good heat resistance, be suitable for hot curing and have a relatively low viscosity at curing temperatures to give good flow in the mould.

About one part of resin, containing $0 \cdot 25\%$ of benzoyl peroxide as curing catalyst, is mixed with three parts of fillers, such as chalk or china clay, in a Werner–Pfleiderer-type mixer. A small quantity of glass fibre is sometimes incorporated in the mix and, if so, this is added near the end of the mixing cycle to minimize damage to the fibre. Occasionally, also, a small amount of an inhibitor such as benzoquinone is added to improve the shelf life of the product.

The products are doughy in consistency and can be cured in moulds at 200 lb/in^2 and $140°C$. They have some outlets for production of car heater covers, battery boxes and similar products where the extra strength given by the glass fibre is essential.

Alkyd Moulding Powders

There is some confusion in nomenclature for these products and the so-called alkyd moulding powders of commerce are usually based on diallyl phthalate. They are made by first preparing a prepolymer from diallyl phthalate by reacting the ester with $0 \cdot 5$–$2 \cdot 0\%$ of benzoyl peroxide at about $80°C$ for several hours; about 25% of the ester is converted to a prepolymer which is soluble in the monomer and in ketones but insoluble in ethyl alcohol.

The prepolymer is thermoplastic and melts at about $90°C$. It may be converted to a moulding powder by both dry and fusion methods but a solvent method is most commonly used. The prepolymer and about 2% of an organic peroxide are dissolved in acetone and blended in a mixer of the Werner–Pfleiderer or Baker–Perkins type with an equal weight of filler, such as short asbestos fibre, and a trace of a mould release agent. A vacuum is then applied and mixing continued until the product is of a doughy consistency; it is then milled on a two-roll mill, with one hot and one cold roll rotating at different speeds, until the cure has reached the desired stage. The milled resin is then granulated and is ready for use.

The major disadvantage of diallyl phthalate moulding powders is their cost. They have a fast curing cycle, about 1–2 min, in compression moulding, and may be moulded at relatively low pressures around 500 lb/in^2; flow characteristics are excellent and they will take up the most delicate configurations of the mould. The moulded products have excellent electrical properties, good tensile strength and good resistance to acids, alkalis and organic solvents. Water absorption is, however, rather high and impact strength relatively low.

Alkyd moulding powders are made and used in the United Kingdom to the extent of about 1000 t/a. Their main uses are for critical electrical components, such as connectors and terminal strips on printed circuit boards, and in large switchgear.

End-use Breakdown

Unsaturated polyesters have a multiplicity of uses; because of the simple way in which they may be used and their capacity to cure cold without the use of pressure they are ideally suited to small scale manufacture and to extempore "one off" applications. Table 10 gives an estimated breakdown of the end uses which absorbed the U.K. output in 1965.

TABLE 10. END USES OF U.K. OUTPUT
OF UNSATURATED POLYESTERS 1965

	%
Castings	5
Surface coatings	6
Filler compounds	11·5
Transport applications	19
Translucent panels	17
Other building uses	3
Boats	18
Electrical	3
Pipes	8
Miscellaneous	9·5

READING LIST

Glass Reinforced Plastics Technology, by W. S. Penn, Maclaren, 1966.

Fundamental Aspects of Fibre Reinforcing Plastics Compositions, by R. T. Schwartz and H. S. Schwartz, Interscience, New York, 1968.

Polyesters Vol. 2, by Parkyn, Lamb and Clifton, Iliffe Press for the Plastics Institute, London, 1967.

Unsaturated Polyesters, by H. V. Boenig, Elsevier, Amsterdam, 1964.

Glass Reinforced Plastics, edited by Phillip Morgan, Iliffe Press, London, 1961.

CHAPTER 6

Epoxy Resins

THE epoxy resins are condensation products of a dihydroxy phenol and a compound containing an epoxy group; the two compounds used for the major part of commercial epoxy resins are 4,4'-dihydroxydiphenyl propane, also known as diphenylol propane or bisphenol A, and epichlorhydrin. The resins were developed mainly as the result of work by the Swiss firm of C.I.B.A. at Basle and by the firm of Devoe and Raynolds in the United States. Shell took over the Devoe and Raynolds patents and both they and C.I.B.A. commenced manufacture in the United Kingdom in the mid-fifties. Shortly afterwards B.X.L. Plastics Materials Group and Borden Chemicals U.K. Ltd. also began to manufacture the resins. The total production, though still small by plastics standards, was increasing rapidly up to 1965, since when growth appears to have slowed down. This may mean that their penetration of fields held by other resins has stopped and epoxies are now increasing only with the growth of those applications especially suited to them. United Kingdom production figures for the past few years are given in Table 11.

TABLE 11. U.K. PRODUCTION
OF EPOXY RESINS
(figures in 000 t/a)

1962	4·1
1963	4·9
1964	5·8
1965	6·8
1966	6·9
1967	7·1

Epoxy resins are relatively expensive when compared with phenol- and urea-based resins and their use is, therefore, confined to those outlets which will bear their high cost and in which their properties of great strength and toughness combined with excellent resistance to chemical attack can be fully exploited.

RAW MATERIALS

Diphenylol propane (DPP) is made by reacting phenol and acetone together under acid conditions in molar proportions of 2:1. The product, a white crystalline solid, is precipitated by the addition of a light hydrocarbon and is separated and washed in a centrifuge.

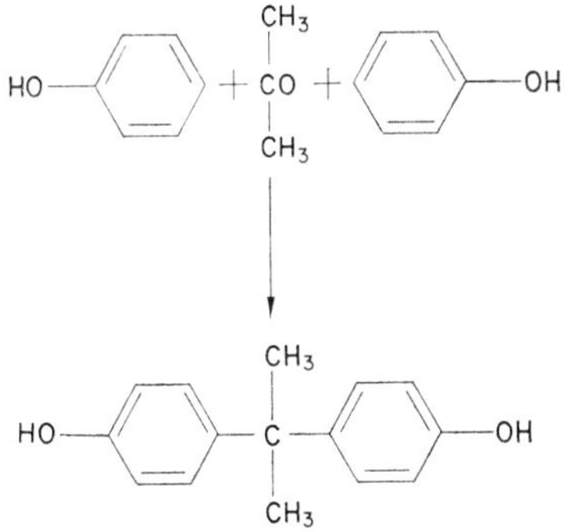

Epichlorhydrin (ECH) is an intermediate in the commercial production of synthetic glycerol by the process, originally devised by the Shell Group, of hot chlorination of propylene to produce monochloropropylene (allyl chloride), followed by reaction with moist chlorine and hydrolysis with milk of lime.

Dichlorhydrin   Epichlorhydrin

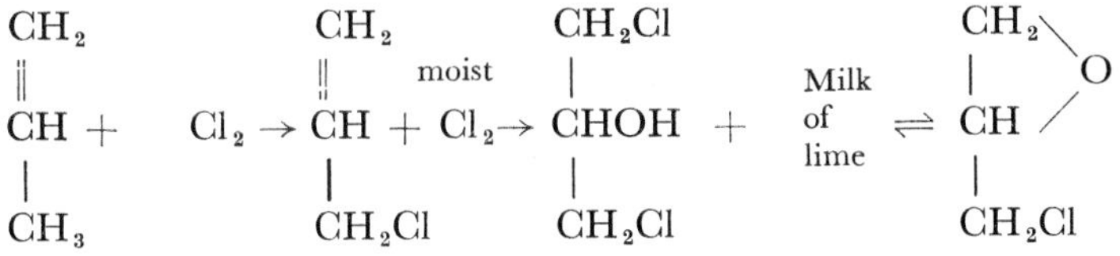

DPP is made in the United Kingdom by Shell but, so far, epichlorhydrin is imported, mainly from Holland.

The two raw materials described above make up the major part of epoxy resin production in the United Kingdom, but other compounds are used in small quantities which may increase in the future. DPP may be replaced by other dihydroxy or poly-hydroxy compounds; for example, a phenol–formaldehyde novolak (p. 62) will react with epichlorhydrin to produce a resin which cures readily and produces a very highly cross-linked product. ECH can also be replaced in whole or in part by other compounds containing an epoxide group; these are usually pro-duced by the hydrogen peroxide or peracetic acid oxidation of compounds containing an olefinic group, such as unsaturated oils and cyclic hydrocarbons, for example cyclohexene. Like the novolak compound referred to above, they are capable of pro-ducing resins with a very high degree of cross-linking which may have higher heat distortion temperatures than the normal DPP–ECH type.

MANUFACTURE OF EPOXY RESINS

The basic reaction between DPP and ECH may be represented as follows:

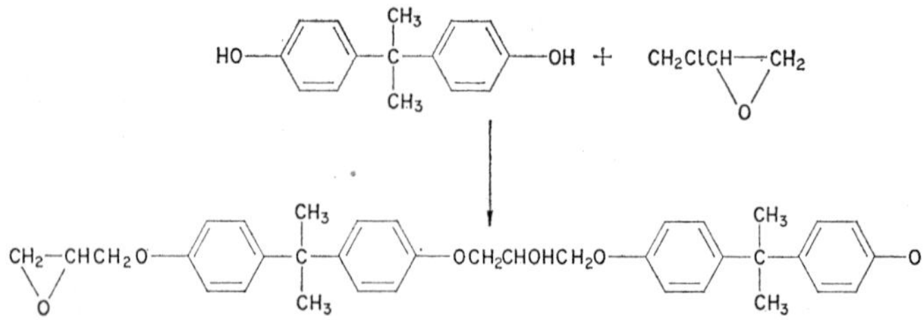

One mole of DPP is reacted, in the presence of caustic soda, with between $1 \cdot 1$ and $2 \cdot 8$ moles of ECH depending on the properties of the resin required. Up to about $1 \cdot 8$ moles of ECH the product is a solid resin but above this proportion the resin is liquid; the lower the proportion of ECH within the range given the higher the molecular weight of the resin, but it appears that a chain length of about twelve units is the maximum.

The resins are produced in equipment basically similar to that described for phenolic resins (p. 68). For solid resins the DPP is dissolved in the minimum quantity of caustic soda at about 35°C and the required amount of ECH is added slowly with vigorous stirring while the temperature is allowed to rise to 65°C. When all of the ECH has reacted, the mix is neutralized and the resin, which separates out, washed repeatedly with hot water to free it from traces of inorganic salts. It is then, while still hot, run off into trays and allowed to set when it forms a clear light brown resin, very similar in appearance to a phenol–formaldehyde novolak. The resin is then ground and is ready for the next stage of application.

For liquid resins the procedure is somewhat different; a typical liquid resin may be made by dissolving the correct proportion of DPP in ECH and heating to 85°C. Caustic soda is now added slowly with stirring and the temperature rise which occurs is controlled by cooling to keep the mixture at about 100°C. An ECH–water azeotrope distils off and is condensed and separated, the ECH being returned continuously to the reaction kettle. Caustic soda to the extent of about a quarter of the weight of the DPP is added in this way and, after the reaction is complete, any unreacted ECH is removed under vacuum. Toluene is now run in to dissolve the resin and the aqueous layer which separates at the bottom of the kettle is run off. The resin solution is now washed, first with aqueous caustic soda to remove any residual chlorine compounds, then with acid sodium phosphate solution to remove any caustic soda and finally with water. The toluene is then distilled off and the resin filtered until clear when it is ready for use.

Curing

Neither the solid nor the liquid resins will cure by heat alone and a cross-linking agent must always be added. The chemical reactions which take place in the curing of epoxy resins start with the opening of the epoxy ring, usually by an amine or a hydroxy group, and the cure proceeds through the further reactions of the active groupings produced in this way, according to the curing agents being used. These reactions are quite complex, not so much in themselves, but because of the number of alternatives which exist, so that the cross-linking mechanism is by no means so simple as it is, for example, with the unsaturated polyesters (p. 97). For a detailed treatment of the cross-linking processes the student is referred to the reading list at the end of the chapter.

The curing agents are usually compounds with two or more amino groups; for the liquid resins, acid anhydrides can be used. Typical amino compounds are paraphenylene diamine, diethylene triamine or a polyamide such as that produced by reacting a dicarboxylic acid with a diamine to form a chain of 5–15 atoms. The cheapest and most commonly used anhydride is phthalic anhydride but hexahydrophthalic anhydride and chlorendic anhydride (also known as HET acid—see p. 105) may also be used. The diamines have a tendency to cause dermatitis and, in order to make them suitable for general use, it is usual for an adduct to be prepared under controlled conditions; this may consist, for example, of equal parts of resin and diamine which will function satisfactorily as a curing agent while the danger that it will cause dermatitis is greatly reduced. The curing agents are required in substantial quantities which may vary from 15 parts per 100 of resin for some amines to 30 parts for phthalic anhydride and even to equal amounts of resin and curing agent for some polyamides.

Some diamines will bring about a cure at room temperature within 2–3 days but most polyamides require much longer and, in some cases, will only cure completely at higher temperatures. Anhydride curing agents always require heat and, in all cases, the cure is effected much more quickly at elevated temperatures. The required conditions may vary from 2 hr at 100°C for the faster curing blends to 6 hr at 180°C for those curing more slowly. In spite of the high temperatures and relatively long times required for a proper cure, the shelf life of resins containing the curing agent is usually rather short and it is necessary to market the resins and curing agents separately.

An entirely different curing technique is to esterify the resin with an unsaturated acid to produce a material which will dry in air like a drying oil (see below under surface coatings).

Compounding

It has already been noted under curing techniques that the curing agent is required in substantial amount and, therefore, forms a significant part of the final product. Thus it acts, to some extent, as a compounding as well as a curing agent and, if cheap enough, may have a significant effect in cheapening the resin formulation which may be still further reduced in cost by the addition of diluents and fillers.

The incorporation of diluents in the resins normally lowers the viscosity and permits a higher proportion of filler to be added. Non-reactive diluents, such as aromatic hydrocarbons, are sometimes used but are not very effective and reactive diluents, which may also be regarded as partial curing agents, are more popular. Typical compounds are epichlorhydrin and various glycidyl ethers produced by reacting alcohols with epichlorhydrin; a polymerizable monomer such as styrene may also be used.

Usually the main advantage of adding a filler to an epoxy resin formulation is a reduction in cost but there may be several other beneficial effects as well. These include reduction of the temperature rise during cure, together with more rapid removal of the heat of reaction, and reduced shrinkage. The filler will also produce significant changes in the physical properties of the cured product; in particular, the coefficient of thermal expansion may be reduced and the thermal conductivity greatly increased. Almost any inert powdered inorganic material may serve as a filler and powdered metal, metal oxides, finely powdered silica and stone and coarse sand have all been used.

APPLICATIONS

Epoxy resins have outstanding qualities of toughness, electrical insulating power and adhesion to many surfaces combined with excellent resistance to attack by chemicals. They have fairly high heat distortion temperatures and retain their desirable properties over a relatively wide temperature range. Typical properties are shown in Table 12.

TABLE 12. PHYSICAL PROPERTIES OF EPOXY RESINS

Compressive strength	15–30,000 lb/in²
Power factor (10^6 cycles)	0·01–0·02
Dielectric constant (10^6 cycles)	3·5–4·0
Dielectric strength (volts/mil)	350–450
Flexural strength	10–20,000 lb/in²
Heat distortion temperature	up to 300°C
Tensile strength	5–12,000 lb/in²
Density	1·2

Although the proportion is declining, more than half the total output of epoxy resins is still used in surface coating applications; these are outside the scope of this book and only brief details are given in the section below.

Surface Coating

There are three important methods of using epoxies in surface coating applications and these are briefly described. Suggestions for further reading are given at the end of the chapter.

Esterification with Drying Oil Fatty Acids

A medium molecular weight solid epoxy resin may be reacted with about two-thirds its weight of an unsaturated fatty acid, such as linoleic acid, but a variety of formulae can be used to give resins with different properties. These may be formulated with other resins, drying oils, solvents and thinners to give both air drying and stoving paints and varnishes for a wide variety of uses.

Combination with Phenolic, Urea and Melamine–Formaldehyde Resins

In this application a high molecular weight solid epoxy resin, dissolved in a solvent such as methyl ethyl ketone or acetone, is used and is combined with the formaldehyde based resin in the proportion of about 70/30, phosphoric acid being added as a catalyst. Paints and varnishes made up with this material may be cured at 150–200°C and combine outstanding flexibility, toughness and adhesion with excellent heat resistance and good electrical properties. They are used as wire enamels, electrical impregnating varnishes and for lining cans, drums and tanks. Urea resins are not quite so good as phenolics in the properties of the product but they are somewhat cheaper and they do allow the formulation of pastel shades of good colour retention; melamine resins give the best properties but are considerably more expensive.

Polyfunctional Amine Cured Systems

A low molecular weight epoxy resin is used and, with diethylene triamine or triethylene tetramine as curing agents, a paint may be formulated which will cure at room temperature. Aromatic amines and polyamides require higher temperatures to effect a cure and must be formulated in stoving compositions; these probably utilize more fully the excellent adhesion and corrosion resistant characteristics of the resins when applied to metals. All direct cured systems need care in formulation; one amine hydrogen atom is required for each resin epoxy group in order to obtain a complete cure. By adding a reactive diluent, such as butyl glycidyl ether, to the resin, solventless varnishes may be formulated.

Electrical Applications

Liquid epoxy resins are used for potting small electrical components and provide both insulation and protection from physical damage. This is essentially a casting application in which a block of resin is cast round the component, care being taken to ensure that the component is perfectly dry and that the resin fully penetrates all its interstices; any required electrical connections must, of course, be led to the surface of the block. A typical potting operation involves the following steps:

1. Assemble the component in a suitable case and load into a vacuum chamber.
2. Bake for 6 hr at 130–170°C.
3. Evacuate for 6–9 hr at 1 mm pressure.
4. Flood the case with liquid resin containing the curing agent.
5. Admit air to force resin into all interstices of the component.
6. Drain off excess resin.
7. Cure in an oven over a period of several hours, starting at a low temperature and increasing it as the cure proceeds.
8. Cool, remove the component from the case and trim.

Phthalic anhydride is a suitable and cheap curing agent; 40% on the weight of the resin is required. The heat distortion temperature of the resin, at 75°C, is relatively low for an epoxy; much higher heat distortion temperatures can be obtained by using 10% of chlorendic anhydride or 17% of metaphenylene diamine; the latter also gives a much higher impact strength. Both formulations, however, are much more expensive than that using phthalic anhydride.

An important consideration in potting is the heat developed during cure, which may damage the component unless controlled. Design of the component being potted, resin formulation and rate of cure are all relevant factors to be considered when evolving the potting procedure. Fillers such as powdered mica may be added to potting compositions and will both reduce shrinkage and improve thermal conductivity; they cannot, however, be used when the resin is required to impregnate a coil winding or similar irregular surface.

Adhesives

It was probably in the form of sticks of adhesive that epoxy resins first became available to the general public. The combination of polar groups, epoxide, hydroxyl and, from the catalysts, amine groups gives excellent adhesion to metals, glass and ceramics. When properly cured the cohesive strength is such that failure often occurs in the materials being bonded rather than at the resin interface and this has permitted some quite spectacular applications of the resins in structural work.

The fact that the resins cure without elimination of water and that there is no volatile solvent present, makes it possible to bond two surfaces with no more than contact pressure; the bond is improved by the very low shrinkage on cure shown by epoxies, far less than with other synthetic resin adhesives. The excellent resistance of the resins to water and solvents ensures a durable bond. Adhesives are frequently formulated with up to 50% of filler, calculated on the resin content; typical fillers are alumina, talc or mica and these actually increase the shear strength of the bond while shrinkage is still further reduced. As curing agents, aromatic diamines such as metaphenylene diamine give the best heat resistance and polyamides the best flexibility and toughness; both kinds normally require heat for a proper cure.

Most epoxy adhesives are supplied as two-part systems, the liquid resin and the curing agent. Some curing agents, such as the adduct of boron trifluoride and diethylamine, do not react at all in the cold and may be incorporated in the resin to give a one component system. Cold curing adhesives can be formulated by mixing the resin with 10% of triethylene tetramine or tetraethylene pentamine immediately before application.

Press Tools

This is a most important application of epoxy resins; filled epoxies are cheaper, on a volume basis, than most suitable metals, while the labour costs of making the tool are only a fraction of those for metal tools. Tool making is generally a casting process and begins with the production of a pattern in wood or plaster which must be thoroughly dry. The pattern is then coated with grease or wax and sprayed with polyvinyl alcohol to ensure that the tool can be released from the mould. A low molecular weight liquid resin is chosen and is mixed with approximately an equal weight of filler; this may be calcium carbonate for easy machining, graphite for good abrasion resistance or iron oxide for maximum heat resistance, according to the use for which the tool is required. The curing agent, preferably an aromatic diamine for maximum hardness, is added and the mixture is poured into the mould, two pours being necessary for large castings.

Epoxy resins can be used for making templates, gauges, jigs and fixtures of all kinds and are even used for lining drop hammers.

Reinforced Plastics

Epoxy resins can be used in exactly the same way as unsaturated polyesters to manufacture structures of all kinds; 15% of metaphenylene diamine is an excellent curing agent for this purpose. Epoxy glass fibre products are both stronger and more resistant to chemicals than the corresponding polyester glass fibre materials; they are, in consequence, widely used for pipes in the petroleum and chemical industries and are especially useful since a properly made pipe will stand pressures of 1000 lb/in^2.

Reinforced plastics based on epoxies are, however, appreciably more costly than those from unsaturated polyesters and their use will be confined to those applications where their superior properties are essential.

Future of Epoxy Resins

Up to the present, epoxies have grown by penetration of fields of use held by other resins where their superior properties have justified the increased cost. This process appears to have come to an end and output is now growing only with the growth of the specialized applications. In the absence of any unexpected development of new uses, therefore, output can be expected to grow relatively slowly. Some reduction in cost may be possible, but the raw materials for epoxy resins are sophisticated chemicals and it cannot be anticipated that they will ever rival the simpler plastics materials in cost.

READING LIST

Epoxy Resins, by Irving Skeist, Reinhold, New York, 1960.
Epoxy Resins, by Lee and Neville, McGraw-Hill, New York, 1957.
Electronic Packaging with Resins, by Harper, McGraw-Hill, New York, 1961.

CHAPTER 7

Polyurethanes

THE fundamental reaction of an isocyanate with an alcohol to form a urethane was discovered by Wurtz in 1848 but it was nearly a century before the possibilities for the production of high polymers by reacting di-isocyanates with diols were realized. The first commercial success was achieved by the Bayer Company in Germany; the driving force was, apparently, the dominating patent position of the DuPont Company in the nylon field following the successful fundamental researches of Carothers. This led Bayer to seek another route to similar materials and they found that the reaction between hexamethylene di-isocyanate and 1,4-butane diol gave a polymer which could be drawn into high tenacity fibres (Perlon U). Later Bayer extended the reaction to the production of resins for surface coatings and adhesives and to the production of elastomers; they were certainly aware of the possibility of producing foams by adding extra isocyanate and water but the first bulk production of polyurethane foams was carried out in the United States. The first production in the United Kingdom was in the mid-fifties by Kay Bros., under licence from Bayer, and they were quickly followed by Dunlop and other manufacturers.

Although, at first, all raw materials were imported and Bayer dominated the field through its licensees, it was not long before British companies, such as Shell Chemicals and Lankro and the British branches of American manufacturers, for example, Union Carbide, Dow and Pfizer, began to produce a range of diols and triols for the new industry. I.C.I. also began to produce isocyanates and to develop their own processes for polymer production;

they were joined a little later by DuPont (U.K.) Ltd. Polyurethane manufacture is now well established in all industrialized countries and is growing rapidly. Growth seems likely to continue for some time yet and there is no reason to believe that the possibilities for further development have been exhausted.

CHEMISTRY OF THE POLYMERS

As with the epoxy resins described in the previous chapter, the chemistry of polyurethanes is made complicated by the existence of a large number of alternative modes of reaction so that a detailed treatment is not possible here; suggestions for further reading are given at the end of the chapter. The urethanes are derivatives of carbamic acid NH_2COOH; the acid itself does not exist in a free state but ethyl carbamate, or urethane, may be produced by indirect methods. The general reaction between isocyanates and alcohols to form urethanes may be written:

$$RN{=}C{=}O + R'OH \longrightarrow RNHCOR' \atop \displaystyle \| \atop \displaystyle O$$

The central grouping $(NHCO_2)$ may be regarded as the urethane group and a polyurethane will contain a multiplicity of such groups. This general reaction opens up a vast number of possibilities since there is virtually no limitation on the type of groups represented by R and R'. Three other reactions of the isocyanate group are given below as they are of major importance in polyurethane chemistry but they by no means exhaust the possibilities.

With carboxyl groups isocyanates give an acid amide type of compound and carbon dioxide is eliminated.

$$RN{=}C{=}O + R'COOH \longrightarrow R'CONHR + CO_2.$$

With an amine the product is a substituted urea

$$RN{=}C{=}O + R'NH_2 \longrightarrow RNHCONHR'.$$

With water the isocyanate group hydrolyses to an amine and carbon dioxide; the amine may immediately react with more isocyanate

$$RN{=}C{=}O + H_2O \longrightarrow RNH_2 + CO_2.$$

It should be noted that the active hydrogen entering into the primary reaction is not eliminated but is transferred to the nitrogen atom of the isocyanate group; it is still active, although less so than before, and may take part in further reactions.

As already described for unsaturated polyesters and epoxies, the reaction between a di-isocyanate and a diol produces a long chain polymer. These chains may cross-link slowly through the action of further isocyanate on the hydrogen of the urethane grouping but the use of a triol will ensure more rapid and more efficient cross-linking. The properties of the final product depend in the first place on the molecular structure and chemical properties of the reactants although formulation of the mix and techniques of polymer production also play a very important part.

When the reaction between isocyanate and hydroxyl groups is the only one taking place, the products may range in flexibility from hard brittle solids to elastomeric materials according to the reactants used. Thus, the earliest applications covered the use of these materials as fibres, surface coatings and rubbers. It was quickly found that if the mix contained free carboxyl groups or water, the carbon dioxide generated from the reaction of these materials with isocyanate could cause the polymer to be produced in the form of a foam. Although they still have important and growing applications in the field of surface coatings and, to a lesser extent, as adhesives and rubbers, it is in the production of foams that the greatest commercial development has been achieved. As with the unfoamed polymers, the foams may vary from hard, rigid, closed-cell materials to open cell foams of great flexibility and resilience. The general principles may be simply stated: reactants with a branched structure in a highly cross-linked polymer will give a hard, rigid product whereas elastomers need reactants with long unbranched chains, preferably aliphatic, in a polymer with relatively few cross-links. When a foam is required the reaction mixture must contain some free water and/or some reactants with free carboxyl groups; in some applications, however, the production of the foam by the addition of a suitable blowing agent, such as a fluorocarbon, to the mix rather than dependence solely on evolved carbon dioxide is becoming the accepted practice.

RAW MATERIALS

From the general nature of the reactions described in the preceding section it is clear that there is potentially a large range of raw materials suitable for the production of polyurethanes. Practical consideration, however, demand a cost–performance relationship which is possessed by a relatively small number of those compounds potentially available; the more important of these are described below.

Only two di-isocyanates have achieved large scale industrial use for polyurethane manufacture. These are tolylene di-isocyanate and 4,4-di-isocyanatodiphenyl methane, commonly known as TDI and MDI respectively. Several reactions for production of isocyanates are described in the textbooks but the only one of industrial importance is the reaction of phosgene with primary amines. For tolylene di-isocyanate the reaction takes place in two stages for which the equations may be written:

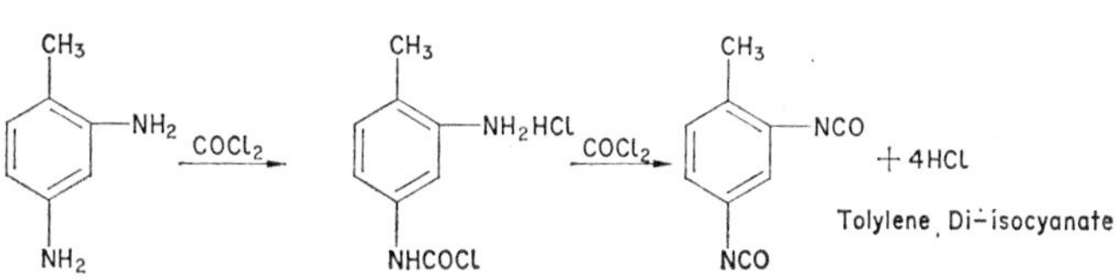

need reactants with long unbranched chains, preferably aliphatic, in a polymer with relatively few cross-links. When a foam is

There are two U.K. manufacturers of di-isocyanates; I.C.I. produce both TDI and MDI at a plant near Fleetwood, Lancs. and DuPont (U.K.) Ltd. make TDI at their plant near Londonderry in Northern Ireland.

For the hydroxyl-containing constituent a wide choice of compounds is available on an industrial scale; one of the first compounds employed in Germany was 1,4-butane diol, possibly because this was a readily available intermediate from the German synthetic rubber programme. Currently, however, it is customary to use a hydroxy compound, loosely called a polyol, of much higher molecular weight and containing more than two hydroxyl groups per molecule. Two main types of compound are in use: the polyesters and the polyethers. The polyester may be an aliphatic compound made from ethylene or propylene glycol and adipic or sebacic acid in molecular proportions chosen so that the polyester has terminal hydroxyl groups. An alternative is to use an alkyd (see p. 99); this opens up tremendous possibilities for the production of polyhydroxy materials capable of forming extensively cross-linked resins. In either case it is clearly possible to formulate so that there are free carboxyl groups left in the polyester, thus promoting the formation of acid amide-type linkages. The use of amines as catalysts opens up a further series of reactions which add to the complexity of the resin.

While the polyesters have proved satisfactory for rigid foams, flexible foams made with them tend to show high hysteresis— that is to say they recover their shape rather slowly after deformation under load. It was found that polyethers with terminal hydroxyl groups were much superior in this respect and could be formulated to give a foam of excellent quality. The first compounds used were polymethylene derivatives made by converting furfuraldehyde to tetrahydrofuran followed by opening of the furan ring but these were soon supplanted by polymers of ethylene and propylene oxide, a small amount of water being added to provide the terminal hydroxyl groups. The derivatives of propylene oxide are now generally preferred.

$$H_2O + nCH_2 \!-\! \underset{\underset{\displaystyle O}{\diagdown \diagup}}{CHCH_3} \longrightarrow H(OCHCH_2)_n OH$$
$$\underset{\displaystyle CH_3}{|}$$

These polyglycols contain only two OH groups per molecule

but, by starting a reaction with a polyhydroxy compound, such as glycerol, instead of water a polyether chain may be built up on each hydroxyl group of the parent compound until the desired molecular weight is attained and each chain will have a terminal OH group. Alternatively a straight polypropylene glycol may be mixed with glycerol, or a similar chain branching agent, during the preparation of a prepolymer as described below.

POLYMER PRODUCTION

Most of the polymerization and cross-linking reactions described take place quite rapidly at normal temperatures and are exothermic. Production problems, therefore, are very largely concerned with rapid and efficient mixing of reactants, with dissipation of reaction heat and with control of the various reaction rates so that the reactions take place in proper sequence. In the production of a foam, for example, it is of no use to have the CO_2-producing reactions proceed so rapidly that they outrun the formation and cross-linking of the polymer.

Flexible Foams

Two basic methods of flexible foam manufacture, generally called the prepolymer and the one-shot processes, are in common use but the latter is gradually displacing the former.

In a typical example of the prepolymer process a polypropylene glycol, molecular weight about 2000, is mixed with glycerol in 3:1 molar proportions. The mixture is heated to 80°C with stirring and about 1·5 moles of TDI are added; an exothermic reaction takes place and the mixture increases in viscosity. The temperature is kept below 80°C by cooling and, when a viscosity of 3–4 stokes at 80°C has been reached, excess TDI is added and the reaction allowed to continue until no further increase in viscosity takes place. This product is the prepolymer and, for storage purposes, it is usually stabilized by adding a trace of benzoyl chloride.

Water may be used as the chain branching agent instead of a triol; only about 0.3% by weight on the polypropylene glycol is required and the rate of viscosity increase must be carefully controlled to allow any liberated CO_2 to escape without, at this stage, causing foaming.

The prepolymer is converted to a foam by mixing it with about 3% of water, when the viscosity increases very rapidly so that the CO_2 evolved cannot escape and causes the mixture to swell into a foam. Due to the excess of TDI used in its manufacture, the prepolymer will have terminal NCO groups which, on reaction with water, give CO_2 and NH_2 groups. The NH_2 groups will react rapidly with further NCO groups to give cross-linking. Regulation at this stage is attained by mixing a small amount of a silicone oil with the prepolymer and a tertiary amine with the water. The silicone oil controls the size of CO_2 bubble produced and thus stabilizes the foam and gives it an even texture while the tertiary amine can react with NCO groups to prevent cross-linking.

The mixing may be carried out batchwise in suitable equipment and the mixture, as it starts to foam, poured into a mould or the reactants may be mixed in a mixing head and the mixture discharged continuously on to a conveyor belt to form a moving block of foam which can be cut into suitable lengths as it comes from the machine. In all methods the newly formed foam is passed through a heating zone to remove any residual tackiness, then through rollers to break up any closed cells and is finally heated to 70°C to complete the curing process.

The One-shot Process

The main difference between this and the prepolymer process is that the polypropylene glycol is combined with the chain branching agent during its production (see p. 139) and is thus ready for use. The foam-making machine has a continuous mixing head which is provided with metering equipment for all the ingredients. The mixture is discharged into moulds passing through the machine on a conveyor belt or direct on to a conveyor belt for production of continuous slabstock.

A typical formulation would be: propylene oxide–glycerol condensate (molecular weight about 3000) 100 parts; tolylene di-isocyanate 40 parts; water 3 parts; silicone oils 1 part; dibutyl tin dilaurate 0·04 part; and an amine (e.g. N-methyl morpholine, triethylene diamine or dimethylethanolamine) 0·5 part. The water, amine and silicone are usually premixed to avoid more than four streams to the mixing head. As in the prepolymer process the amine controls the amount of cross-linking while the tin compound limits the degree of polymerization. Considerable modifications may be made in the properties of the foam by adjusting the proportions of the reactants.

Polyester foams can be made by analogous methods; the polyester for flexible foams is normally the reaction product of a trihydroxy alcohol and a long chain dicarboxylic acid.

Properties of Flexible Foams

The foams have open cells and can vary widely in density but about 2 lb/ft^3 is the usual figure. This is considerably lighter than most foams made from rubber latex which, with roughly the same properties, weigh 3·5 lb/ft^3. At the same density polyurethane foams can bear a heavier load and are stronger than the rubber foams, both in tensile and tear strength. They have good resistance to rot, vermin and water and to some organic solvents but not, for example, to acetone. The foams bond well to textiles, glass, wood and metal, are less flammable than rubber and can be made self-extinguishing. Their permanent set after compression is nearly, but not quite, as good as that of rubber latex foams and it is in this property that the difference between polyester and polyether foams is most marked. The former are much stiffer at low loads, almost collapse under high loads and show a very poor recovery after high loading. Polyester foams have slightly higher mechanical strength than polyether foams but this does not compensate for their poor load characteristics.

Uses of Flexible Foams

As might be expected, polyurethane foams have found similar uses to rubber latex foams, which they have to some extent replaced, in automobile seating, upholstery, cushioning and mattresses. Because of the ease with which they can be bonded to textiles they have found applications as shoulder pads, interlinings, draught excluders, carpet backings and underlays. Their open cell structure makes them useful as sponges. About 10% of the 1967 consumption was of polyester foams and these are mainly used for textile laminating. The estimated end-use breakdown was as shown in Table 13.

TABLE 13. U.K. END-USE BREAKDOWN
FOR FLEXIBLE POLYURETHANE FOAMS 1967

	tons
Furniture	9500
Automobile seats, etc.	7500
Other household uses	4000
Bedding, mattresses	2000
Foam backings	2000
Other uses	1000

Rigid Foams

These are similar to flexible foams in overall composition, the main difference being that the components are chosen to give a much greater degree of cross linking, thus ensuring a rigid structure and closed instead of open cells. For example, a propylene oxide–glycerol condensation product of molecular weight around 300 instead of 3000 would be chosen; this would have many more cross-linking points per unit weight of polyether and would favour cross-linking at the expense of molecular weight growth. The use of MDI instead of TDI is also advantageous in this respect.

Excellent rigid foam has been obtained by reducing the proportion of water and, hence, the expensive isocyanate and incorporating trichlorofluoromethane (boiling point $23 \cdot 8°C$) as a blowing agent. While blocks and slabs can be produced by methods similar to those for flexible foams, a property of great importance is that the compositions can be foamed *in situ* so that irregularly shaped cavities can be filled. In this way the foams may be used as insulating material for refrigerated vessels of all types, in transport vehicles, in buildings and even in ships. Using the one-shot process there is usually a delay of 20–30 sec after mixing before foaming starts. Thus, by using a mixing head of large capacity, e.g. up to 70 lb/min, quite large cavities can be filled.

Properties and Uses of Rigid Foams

As implied in the previous paragraph, the main use of rigid polyurethane foam is for insulation, especially low temperature insulation down to $-60°C$, although the foams will stand up to temperatures of $100°C$, and even higher temperatures when specially formulated. Their closed cell structure gives them excellent resistance to water absorption and water vapour transmission. Most rigid foams are made with densities of $1 \cdot 5$–3 lb/ft³. In this density range the product has a compressive strength of 25–40 lb/in², and a modulus of elasticity from $250/1000$ lb/in². There is some evidence that the conventional carbon dioxide foamed polyurethane has slightly better mechanical properties than that blown by trichlorofluoromethane.

A very large potential outlet for the material is in the production of sandwich board for partitions in buildings. The insulating properties of such board against heat are particularly advantageous.

To some extent, polyurethane foam competes directly with foamed polystyrene. Polystyrene is probably slightly cheaper per unit of insulating power but is more difficult to apply as it cannot be formed *in situ*. Polyurethane is superior at very low temperatures and, as a thermosetting material, is much more heat resistant; it is destroyed by fire but retains its rigid structure longer than polystyrene foam. Rigid polyurethane foams also have the advantage that they adhere strongly to the surfaces on which they are formed and, in some circumstances, the compositions may be sprayed direct on to the surfaces to be insulated.

OTHER USES FOR POLYURETHANES

Although now predominantly used for foams these materials, as noted in the introduction to this chapter, were originally used in the more conventional applications for polymers. Their early uses in surface coatings and adhesives were overshadowed in the 1950's by epoxy resins but applications in these and other fields are now growing on their own merits.

Many polyurethane paints are now on the market, both one-can and two-can systems. Considerable skill and experience is required in their formulation to give products of good shelf life; both stoving and air-drying systems are available. As with epoxy resins, adhesion and chemical resistance are the important properties which polyurethanes give to paints. They are particularly good as wood finishes, especially for wood floors, since they give great surface hardness and scuff resistance.

Excellent adhesives can be formulated from di-isocyanates and glycols, with or without fillers, which cure rapidly on application of heat. By blending with a rubber cement an excellent alternative to resorcinol formaldehyde latex may be produced for bonding rubber to synthetic fibres in the production of reinforced rubber products such as belting and tyres.

By careful formulation it is possible to produce polyurethanes suitable for practically all of the conventional plastics applications and moulding powders, extrusion and injection compounds and the like may be made as required. Formulations similar to those for foam production are used but with exclusion of water; thermoplastic or thermosetting compounds can be obtained according to the formulation. The thermoplastic material closely resembles nylon in properties but is much more expensive. As with epoxy resins, cost is likely to limit the use of these materials to those highly specialized applications where the extra expense is justified.

Elastomers

Polyurethanes can form synthetic rubbers with outstanding properties which have higher tensile strength than any other known rubber, have very high resistance to abrasion and are practically untearable. They are made in modest quantities by DuPont and Goodyear in the United States, Bayer in Germany and I.C.I. in the United Kingdom. A wide range of formulations is available but the basic principle seems to be to react a long chain polyester, such as the adipic ester of a polyethylene glycol, with just sufficient di-isocyanate to leave some free hydroxyl groups. Compounding is done on heated rollers in the usual way with fillers, softeners, etc., and further di-isocyanate is added; in order to limit toxicity this must be as high boiling as possible and MDI or naphthalene 1,5-di-isocyanate have been used. The compounded mixture can then be heat cured. More recently formulations have been developed to cure with peroxides or, by arranging that residual unsaturation is present in the rubber, with sulphur as for natural rubber.

The cost of these formulations has, however, restricted their use to special applications, but the possibility of their being used to make tyres which would outlast the life of the car to which they were fitted is interesting in view of the current public concern with tyres as a possible cause of road accidents. So far, however, any tyres made have suffered from a tendency to melt on skidding and are not really a practical proposition for cars and lorries. At present uses for polyurethane elastomers include manufacture of printers' rollers, tyres for fork lift trucks, anti-vibration mountings, tips for ladies' stiletto heels and similar specialized small scale outlets.

SCOPE FOR FUTURE DEVELOPMENT

Polyurethanes are still relatively new among plastics materials and their chemical composition permits a wide range of raw materials to be used in their manufacture. Many possibilities for new formulations and for new uses still remain to be explored and development is likely to continue for some years ahead. Their cost is still high and is likely to remain so relative to the giants among plastics materials but their performance is so good that, in some areas, a relatively modest reduction in cost might considerably extend their field of application.

READING LIST

Polyurethanes, by B. A. Dombrow, Reinhold, New York, 1957.
Polyurethane Foams, edited by T. T. Healey, Iliffe Press, London, 1964.
Polyurethane Chemistry and Technology, by J. H. Sanders and K. C. Frisch, Interscience, New York, 1962.

PART III

Thermoplastics

CHAPTER 8

Polyolefins

IT WAS noted during the discussion of functionality on p. 120 that all mono-olefins are difunctional and will, under the right conditions, form long chain polymers. Only a few of them have gained commercial acceptance but the simplest of them all, polyethylene, has, in the 30 years since it was first produced commercially on a pilot plant scale, become one of the major thermoplastics throughout the world. It will, together with polypropylene, the next most important polyolefin, form most of the subject matter of this chapter; brief descriptions of some copolymers of ethylene and propylene, both together and with other olefins, and some homopolymers of higher olefins are also included. The polymers of styrene, which can be regarded as phenylethylene, and of vinyl chloride or chloroethylene, which are major thermoplastics in their own right, are dealt with in the next two chapters.

There are two kinds of polyethylene, which may be differentiated by their method of production or by their density. The first to be discovered is produced by a high pressure process and is of relatively low density, whereas the other is produced by a low pressure process, which was discovered much later, and is of relatively high density. In this chapter they will be referred to as low density and high density polyethylene respectively.

RAW MATERIAL

Ethylene, the simplest olefin, was first isolated at the end of the eighteenth century by dehydration of ethyl alcohol (ethanol). The study of the reactions of this ethylene, or olefiant gas, made some contribution to the growth in knowledge of organic chemistry which took place in the nineteenth century but ethanol from fermentation processes remained the source of supply until the advent of cracking plants in U.S. oil refineries during the early years of the twentieth century. The first commercial uses of ethylene from these sources were for the production of synthetic ethanol and ethylene oxide during the 1920's. Both of these could be produced from ethylene streams containing paraffinic diluents; the large scale manufacture of ethylene in the purity required for production of polyethylene did not become general until after the Second World War. Even the first commercial manufacture of polyethylene was based on ethylene from ethanol. Now commercial ethylene is produced by cracking natural gas or light liquid fractions from petroleum in plants specially designed for the purpose. This change of basic raw material, together with recent important improvements in the efficiency of cracking and gas separation plants for ethylene production and the development of new methods of transporting the gas, either by pipeline or as a liquid in bulk, have made possible the dramatic reductions in price of polyethylene which have taken place during the last 10 years.

Ethylene for polymerization has to be of high purity—at least $99 \cdot 8\%$—and, in particular, the acetylene content must be no more than a few parts per million. Propylene for polypropylene is generally co-produced with ethylene in the same cracking plants and it too must be very pure with only a negligible content of methyl acetylene $CH_3C \equiv CH$.

The butanes which have some importance as monomers for commercial manufacture of polymers, are made by fractionation of mixed C_4 streams arising as by-products in commercial ethylene cracking plants and general oil refinery operations.

4-Methyl pentene, which is just beginning to be commercially important, and which may have an interesting future, is made by dimerization of propylene.

$$2CH_2=CHCH_3 \longrightarrow CH_2=CHCH_2CHCH_3$$
$$| $$
$$CH_3$$

LOW DENSITY POLYETHYLENE

This is one of the three important plastics materials which are British inventions, the other two being nitrocellulose and the saturated polyester generally known as "Terylene" (see p. 220). The discovery of low density polyethylene is a fascinating example of the way in which a valuable commercial invention may come out of a purely theoretical investigation.

In 1933 R. O. Gibson was one of a team in the I.C.I. (Alkali Division) research laboratory at Northwich working on the effect of very high pressure on chemical reactions, without the intention of producing any specific product. Heating ethylene and benzaldehyde at 170°C and 1300 atm pressure, he recorded in his notebook the production of a thin layer of a waxy solid. It contained no oxygen and was, therefore, believed to be a polymer of ethylene. An attempt to repeat the experiment with ethylene alone brought an explosive reaction with production of amorphous carbon and some damage to the apparatus.

By 1935 stronger equipment was available and M. W. Perrin and E. G. Williams repeated the experiment at 30,000 lb/in^2 when a white powder was obtained. It now appears that, by a fortunate chance, ethylene added during the experiment contained just sufficient oxygen to give the catalytic effect required.

A detailed research programme, led by E. W. Fawcett, now followed in which the best conditions for the production of the polymer were worked out; a pilot plant was set up in 1938 and, by the end of the year, 1 ton of polyethylene had been produced. Its potentiality as an electrical insulating material was recognized and was fully confirmed by a large-scale evaluation carried out by the Telegraph Construction and Maintenance Company. A small, high pressure plant was erected by I.C.I. and came on stream on 1 September 1939. During the war its use was confined to military objectives, such as high-frequency insulation in Radar, but after 1945 its value for other purposes began to be recognized and consumption grew rapidly. It was first made in the United

States in 1943 but was not produced commercially in Germany until 1950.

United Kingdom produced material has been exported all over the world and, even now, exports are still growing and reached a level of 76,000 tons in 1967. The production–consumption figures for low density polyethylene in the United Kingdom are shown in Table 14. It should be noted that, due to complications introduced by imports and re-exports, the difference between production and consumption does not give the true figure for exports.

TABLE 14. U.K. PRODUCTION AND CONSUMPTION
OF LOW DENSITY POLYETHYLENE
(figures are 000 tons)

Year	Production	Consumption
1962	140	90
1963	161	106
1964	198	124
1965	202	139
1966	219	157
1967	236	175

At first I.C.I. were the only U.K. producers and they are still by far the largest; other companies, however, set up manufacturing plants in Europe and the United States under licence from I.C.I. and these companies soon began to develop their own information on production and processing of the polymer. As the I.C.I. patents began to expire some of these licensees were able to grant licences for the know-how which they had accumulated. Under these conditions three other companies set up manufacturing facilities in the United Kingdom, B.X.L. Plastics Materials Group Ltd. using Union Carbide know-how and Monsanto and Shell using B.A.S.F. know-how. Estimated capacities at the present time (1968) are shown in Table 15.

TABLE 15. ESTIMATED CAPACITY OF U.K. POLYETHYLENE PLANTS

Company	Location	Capacity (000 t/a)
I.C.I. Ltd.	Wilton, Yorks	180
B.X.L. Plastics Materials Group Ltd.	Grangemouth, Scotland	80
Shell Chemicals U.K. Ltd.	Carrington, Cheshire	60
Monsanto Chemicals Ltd.	Hythe, Southampton	50
Total		370

I.C.I. and Shell make their own ethylene but B.X.L. and Monsanto get theirs from the adjoining cracking plants of B.P. and Esso respectively.

Production of Low Density Polyethylene

Chemically, the polymerization of ethylene could hardly be simpler but carrying it out on a large scale poses some very difficult chemical engineering problems. Huge compressors are necessary to compress the large tonnages of gas in circulation to the operating pressure of 2000 atm and the separation of the semi-liquid polyethylene from unreacted gas, followed by re-cycling the unconverted ethylene, with purification if necessary, have all needed the development of chemical engineering techniques to a high level.

The process most widely used in the United Kingdom is little more than a refinement of the process first used by I.C.I.; it consists, basically, of compressing ethylene to 2000 atm and heating it to 170°C in the presence of traces of oxygen. Under these conditions the ethylene is still a gas but its density, about 0·5, is little below its liquid density and the reaction is not dissimilar from a liquid phase reaction. At the reaction temperature and pressure ethylene is readily oxidized and it is believed that the oxygen catalyst, about 0·06%, produces free radicals which initiate the ethylene polymerization. Undesirably high molecular weights are reached unless a chain stopper is present and, for this purpose, about 1% of an inert gas, such as propane, is usually added. The time of contact in the reactor is only a few seconds and an ethylene conversion of about 25% is the most economical level. A schematic flow diagram of the I.C.I. process is shown in Fig. 19.

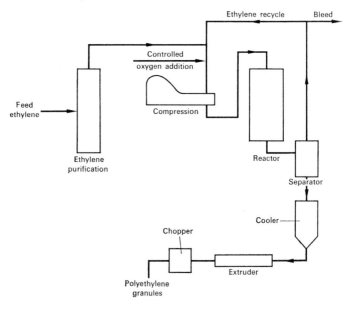

FIG. 19. Flow diagram of I.C.I. process for production of low density polyethylene.

After separation from unconverted ethylene the polymer is worked up in a similar way to other thermoplastics (cf. bulk polymerization of styrene, p. 182). It leaves the separator as a coarse ribbon several inches wide and is passed to the extruder, from which it emerges in the form of thick filaments which are chopped to form granules. The extrusion step can be omitted and the ribbon leaving the separator may be cooled, broken into pieces and ground to a fine powder; this gives a cheaper product and extends the applications where the polymer can be economically used. The powder has found a considerable market in moulding operations.

An intermediate step may be inserted for colouring the product from the reactor before chopping into granules or the granules may be coloured afterwards, e.g. by mixing on hot rolls or in a suitable mixer, followed by a final extrusion into strands and chopping to form coloured granules.

There are several modifications to the original I.C.I. process which are of interest. Union Carbide in the United States (B.X.L. Plastics in the United Kingdom) carry out initial stages very similar to I.C.I. but the separation of the polymer from unreacted gas is rather different. The pressure of the gas–resin mixture is used to force the resin through orifices to form a series of ribbons while the unreacted gas is returned to the reaction chamber.

DuPont use a mixture of ethylene with benzene and a somewhat lower pressure of 1000–1200 atm. The polymer can be condensed out in two stages as the pressure is released and products of different properties are obtained. In the B.A.S.F. process the polymerization is carried out in tubular reactors in contradistinction to the I.C.I. bulk units.

Attempts have been made, notably by I.C.I., to produce a polymer of higher density by the high pressure process. One method has been to carry out the polymerization at relatively low temperatures around 90°C and pressures about 800 atm, in the presence of both a free radical forming catalyst, such as a peroxydicarbonate ester, and a chain transfer agent, typically a halogenated hydrocarbon. Although it reached the brink of commercial success the project was not proceeded with.

HIGH DENSITY POLYETHYLENE

High density polyethylene is a new product and has been on world markets for little more than 10 years. It was first made commercially in Germany, then in the United States, while the first major plant in the United Kingdom came on stream in 1960. When first produced it was considered a great advance on the low density material because of its higher melting point and rigidity and a great future was predicted for it. Then came a period of difficulty and relatively slow growth connected, in the case of Ziegler process material, with problems of catalyst removal. Thus, while low density polymer is produced from high purity ethylene with nothing but a trace of oxygen as catalyst, the high density product requires inorganic catalysts which can leave residues in the polymer and cause deterioration on ageing, particularly in electrical properties.

These difficulties have now been overcome while, in addition, material with enhanced properties has resulted from the co-polymerization of ethylene with a few per cent of propylene or butylene. This has resulted in an astonishing growth rate for production of the high density product, output in the United Kingdom rising from 14,000 t/a in 1961 to 46,000 t/a in 1967.

There are three main processes, Phillips, Ziegler and Standard Oil of Indiana; two of these are established in the United Kingdom, the Phillips process used by B.P. Chemicals (U.K.) Ltd. at Grangemouth with a capacity of 50,000 t/a and the Ziegler process used by Shell Chemicals U.K. Ltd. at Carrington with a gross capacity of 40,000 t/a; some of the Shell plant is dual purpose and is used for polypropylene.

Phillips Process

In this process the solid catalyst and a solvent are present as a slurry and the polymer is held in solution until the catalyst has been removed from the reaction mixture. The ethylene must be free from water, carbon dioxide, sulphur compounds and, contrary to the requirements of the high pressure process, oxygen.

Purified ethylene and cyclohexane as a solvent are charged to the reactor at rates to maintain the ethylene concentration at about 5% by weight. A pressure of 100–500 lb/in^2 is needed and the temperature is maintained at a little over 100°C. The catalyst is chromium trioxide (CrO_3) on a silica–alumina support at a concentration of about $0·5\%$ on the solvent; it is held in suspension either by mechanical agitation or by the lifting power of the incoming feed. The ethylene polymerizes very rapidly, some of the polymer remaining in solution and some becoming entangled with solid particles of catalyst; the latter is withdrawn from the reaction zone as a slurry, the polymer dissolved in additional solvent, filtered to remove catalyst and remixed with the main bulk of the solution. The catalyst can be recirculated but the chromium is gradually reduced to the trivalent state and needs to be reoxidized to the hexavalent state from time to time by air and steam at 600°C.

Any unreacted ethylene is flashed off together with part of the solvent. On cooling, the solid polymer separates and can be centrifuged out, dried and finished by the methods described for other thermoplastics. The solvent plays a very important part in the process in acting as a solution viscosity controller, in dissipating the heat of reaction and, to some extent, in protecting the growing polymer chain from chain breakdown. An outline diagram of the process is shown in Fig. 20. Polymers of molecular weight of about 40,000 are obtained.

A modification of the Phillips process, known as the "Particle Form" process, is to polymerize the ethylene at a pressure of 480 lb/in² in isobutane as a solvent, using a very small amount of chromium oxide catalyst. The polymer is completely insoluble in the isobutane and can be separated very easily. It contains the catalyst but the amount is so small that the polymer is suitable for most applications. Copolymers with butene-1, for example, can be produced by this method.

Ziegler Process

The original process for production of high density polyethylene stemmed from work by Professor Ziegler in Mühlheim. Ziegler was trying to make higher olefins from ethylene and he found

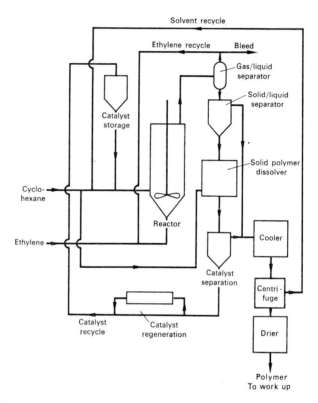

Fɪɢ. 20. Flow diagram of Phillips process for high density polyethylene.

that aluminium alkyls could bring about the polymerization of ethylene to molecular weights of 5000 or so. Each olefin addition step is in competition with a displacement reaction which splits the growing alkyl chain away from the aluminium. Ziegler was trying to promote the displacement reaction to obtain lower molecular weights and he tried the effect of other metal additives. Nickel produced the desired result but titanium caused the reaction to go out of control and Ziegler realized that, when brought under control, he had a method of producing high polymers of ethylene without the high pressures, and consequent chemical engineering problems, of the high pressure process.

The catalyst now normally used is a mixture of titanium tetrachloride and aluminium triethyl; the mode of action of this catalyst is, even now, not fully understood but it is known that the polymerization is brought about by a reaction product of these two reagents, since neither of them alone will cause the production of high polymers of ethylene. The type of reaction which takes place is as follows:

$$(C_2H_5)_3Al + TiCl_4 \longrightarrow (C_2H_5)_2AlCl + TiCl_3 + C_2H_5^+$$

The $C_2H_5^+$ radical then either dimerizes or disproportionates to form ethylene and ethane. The catalyst may be prepared by adding the two components to the polymerization reactor itself or they may be pre-mixed, the precipitated $TiCl_3$ filtered, washed and added to the polymerization to which additional metal alkyl is also added. The second method is found to give better control over catalyst composition. Aluminium alkyls are spontaneously inflammable in air and are decomposed by water so that all operations must be carried out in an inert atmosphere and with exclusion of moisture.

Ethylene or, in a more recent process modification, ethylene containing a few per cent of propylene, is compressed to about 80 lb/in^2 and blown into a suspension of the catalyst in a suitable inert solvent, such as a mixture of C_6–C_8 paraffins at a temperature of about 70°C. The gas stream must be free from sulphur and carbon dioxide and have a very low acetylene content. The polymerization takes place very rapidly and the polymer separates as a slurry. Polymers of very high molecular weight may be produced, even up to 3 million, and some form of control is essential. Precise details of the methods are, of course, process secrets of the manufacturers, but it is known that several factors will limit the molecular weight of the produced polymer. Traces of hydrogen or water tend to restrict the maximum molecular weight attainable and control may also be exercised by varying the ratio of titanium tetrachloride–aluminium triethyl. Oxygen, too, has an important effect and is normally held below $0·05\%$ on the ethylene feed; it reduces the molecular weight of the polymer, increases the rate of reaction and, if too much is present, inactivates the catalyst.

The polymer slurry is first treated to remove the residual catalyst; this is frequently carried out by treatment with dry hydrogen chloride and an alcohol which converts the catalyst into a water soluble form when it can be washed out. The function of the alcohol is to ensure that the polymer particles are thoroughly wetted; without it catalyst removal is incomplete. This treatment leaves the polymer still as a slurry with the hydrocarbon, which is now removed by steam distillation, and the solid

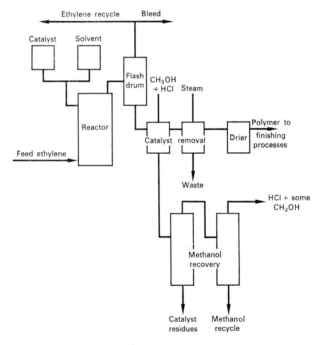

Fig. 21. Flow diagram of Ziegler process for high density polyethylene.

polymer carried in a stream of steam to a cyclone separator where it separates in a relatively dry state.

The rest of the working-up process is similar to that for other thermoplastics and may include colouring, blending with anti-oxidants, light stabilizers and similar materials and ends with a final extrusion into ribbons and chopping into pellets. An outline diagram of the process is shown in Fig. 21.

STRUCTURE OF THE POLYOLEFINS

In order to describe the production of commercial poly-propylene it is necessary to make reference to its structure and it is convenient, therefore, to deal with this in a general way at this point.

In theory the polymerization of ethylene is a simple end-to-end addition of ethylene molecules to form a chain which, when terminated with hydrogen, has the simple structure:

$$CH_3CH_2(CH_2)_nCH_2CH_3$$

A certain amount of chain branching almost always takes place to give a structure:

$$CH_2CH_2CH_2CH_3$$
$$|$$
$$CH_3CHCH_2CH_2CH_2 \quad \text{\textasciitilde\textasciitilde\textasciitilde\textasciitilde} \quad CH_2CHCH_2CH_2CH_3$$
$$|$$
$$CH_3CH_2CH_2CH_2CH_2CH_3$$

The side chains tend to prevent the close approach and orderly packing of the polymer molecules and so increase the flexibility and decrease the softening point and degree of crystal-linity of the product. The sole difference in structure between the two kinds of polyethylene lies in the amount of chain branching which has taken place. The low density material has fairly numerous side branch chains, whereas the high density product has far fewer and that made by the Phillips process is almost completely straight chain. The polymer chains of the high density material can, therefore, pack closely together in an orderly fashion, thus giving a high degree of crystallinity to the product.

An ethylene polymer chain can conveniently be represented on a plane diagram because all the carbon atoms are alike; with polypropylene, however, the case is different. Part of a polypropylene molecule might be represented:

$$
\begin{array}{cccc}
\overset{*}{} & \overset{*}{} & \overset{*}{} & \overset{*}{} \\
-CH_2CH & CH_2CH & CH_2CH & CH_2CH- \\
| & | & | & | \\
CH_3 & CH_3 & CH_3 & CH_3
\end{array}
$$

It can readily be seen that each of the starred carbon atoms is asymmetric and that stereoisomers are possible. With a three-dimensional model it can be demonstrated that, when the asymmetric carbon atoms all have the same configuration, they will arrange themselves in the form of a spiral with the methyl groups on the outside. Three kinds of polymer are then possible:

An isotactic polymer in which the methyl groups all lie on the same side of the carbon chain. This can be represented:

$$
\begin{array}{ccccc}
CH_3 & CH_3 & CH_3 & CH_3 & CH_3 \\
|\quad H & |\quad H & |\quad H & |\quad H & |\quad H \\
C\!\!\diagdown\;| & C\!\!\diagdown\;| & C\!\!\diagdown\;| & C\!\!\diagdown\;| & C\!\!\diagdown\;| \\
/\quad\diagdown C/ & |\quad\diagdown C/ & |\quad\diagdown C/ & |\quad\diagdown C/ & |\quad\diagdown C/ \\
H\quad| & H\quad| & H\quad| & H\quad| & H\quad| \\
H & H & H & H & H
\end{array}
$$

A syndiotactic polymer in which the methyl groups lie alternately on opposite sides of the chain and which may be drawn:

$$
\begin{array}{ccccc}
CH_3 & H & CH_3 & H & CH_3 \\
|\quad H & |\quad H & |\quad H & |\quad H & |\quad H \\
C\!\!\diagdown\;| & C\!\!\diagdown\;| & C\!\!\diagdown\;| & C\!\!\diagdown\;| & C\!\!\diagdown\;| \\
/\quad\diagdown C/ & |\quad\diagdown C/ & |\quad\diagdown C/ & |\quad\diagdown C/ & |\quad\diagdown C/ \\
H\quad| & CH_3\;| & H\quad| & CH_3\;| & H\quad| \\
H & H & H & H & H
\end{array}
$$

An atactic polymer in which the distribution of the methyl groups is random on both sides of the chain.

The spirals of the isotactic polymer will pack closely together, while the less stereoregular molecules of the other polymers will have a more random arrangement. The isotactic material, therefore, has a higher degree of crystallinity, is more rigid, has a higher melting point and greater tensile strength than either the syndiotactic or the atactic polymers.

POLYPROPYLENE

Many of the early attempts to polymerize propylene produced only soft rubbery polymers which had no commercial value, either as plastics or as elastomers. The original Ziegler work was confined to ethylene and attempts to extend it directly to propylene gave only low molecular weight polymers of no particular value. Professor Natta, working in Milan, discovered, however, that by slight modification of the Ziegler catalysts it was possible to produce high molecular weight polymers with a large proportion of isotactic material. Some atactic polymer is always produced but both it and the propylene monomer are soluble in boiling heptane while the isotactic polymer is not. By carrying out the polymerization in a heptane medium it is possible, therefore, to effect a separation of most of the unwanted atactic material.

Propylene for polymerization must be of at least $99 \cdot 5\%$ purity with the remainder mostly propane. It must be free from acetylenes, mercaptans, hydrogen sulphide, oxygen and hydrogen chloride; a trace of hydrogen may, however, be present as a chain stopper to control the molecular weight of the polymer. As with high density polyethylene, it has been found that copolymers of propylene with a small proportion of ethylene have many improved properties and much of the present-day production is the copolymer. The catalyst, normally aluminium diethyl monochloride and titanium trichloride, is suspended in heptane at 70–80°C and propylene is passed in while the suspension is stirred. Most of the isotactic polymer separates out while the atactic material remains in solution.

After removal of any unreacted propylene the slurry is treated with an alcohol, usually isopropanol, and dry hydrogen chloride; this deactivates the catalyst and makes it mainly soluble in the hydrocarbon. The precipitated polymer is separated by filtration or centrifuging and is then subjected to a steam distillation which removes any remaining hydrocarbon solvent and hydrolyses any insoluble products formed in the deactivation of the catalyst; an acid aqueous medium results in which any catalyst residues left dissolve. The polymer is freed from catalyst and water in a centrifuge and is then ready for processing to a saleable form.

Polypropylene readily oxidizes in air and all processing operations must be carried out in an inert atmosphere until antioxidants have been incorporated in it; 0·5% of a phenol and the same quantity of dilauryl dithiopropionate are commonly used. Other additives may be required according to the use to which the polymer is to be put; dyes and pigments may be used for coloured products, light stabilizers are needed for many applications and, for polymer which is to be spun into fibres, dyeing assistants are necessary. The additives can all be blended with the centrifuged and dried polymer which is then extruded into ribbons and chopped to form pellets.

The properties of polypropylene are considerably affected by the proportion of atactic polymer present in the sample. This is not very critical for moulding grades but, for film and fibre grades, the highest possible content of isotactic polymer is required. The separation by differential solubility in the process described above is not, of course, a sharp one and, since economics demand that virtually none of the valuable isotactic material shall be lost, there is always some atactic polymer in the finished product. Much effort has been devoted to minimizing the amount of atactic polymer formed and some of the methods adopted are:

(i) Avoid any aluminium triethyl in the catalyst mix.
(ii) Polymerize at as low a temperature as possible consistent with a reasonable reaction rate.
(iii) Increase reaction rate by increasing the concentration of titanium trichloride and/or by increased pressure.

Production and Producers

Like high density polyethylene, polypropylene was greeted with enthusiasm when it first came on to the market but it, too, has met applicational problems. These have, however, now been overcome and, particularly in injection moulding applications, the recent growth rate has been phenomenal. United Kingdom production figures in Table 16 emphasize this.

TABLE 16. U.K. OUTPUT
OF POLYPROPYLENE
(000 tons)

1962	4
1963	9
1964	16
1965	19
1966	26
1967	48

Properties of the Polyolefins

The three polyolefins have many points of similarity and the important points where they do differ can be more simply emphasized by describing all three together. They are all flexible, waxy translucent materials which can be injection moulded, extruded, cast or blow moulded. They have moderate tensile and high impact strengths and are all resistant to many inorganic acids, but are attacked by organic solvents. They have excellent electrical properties and may all be formed into fibres and films under the proper conditions.

Considerable modification to the properties of all three materials can be made by variations in the methods of production and processing, including, in the low pressure processes, variations in the proportions of the co-monomer, so that it is possible to produce grades of the same polymer which differ more from each other than they do from one of the other polymers. The degree of crystallinity has an important influence on properties and depends both on production and processing methods. Low density polyethylene is usually between 50 and 60% crystalline whereas high density material from the Ziegler process is about 85% and that from the Phillips process 95% crystalline. Crystallinity of polypropylene depends on the amount of atactic, mainly amorphous, polymer present. Although crystallinity is favoured by the isotactic structure, however, even a pure isotactic polymer would not be 100% crystalline. For commercial polymers it is usually between 60 and 70%.

Physical Properties

The important physical properties of all three polyolefins are summarized in Table 17.

The density differences follow the differences in structure and crystallinity already described. With polyethylene, the higher the density the greater the stiffness.

The tensile strengths are not outstandingly high and, as might be expected, are greatest for the more highly crystalline polymers. By biaxial orientation (see p. 172) polypropylene film, and, to a lesser extent film made from polyethylene, can be given tensile strengths very much higher than the values quoted.

The impact strength figures at room temperature follow the density (and stiffness) values; at low temperatures homopolymers of propylene become very brittle. Copolymers largely overcome this disadvantage at the expense of a slight reduction in softening

TABLE 17. PHYSICAL PROPERTIES OF POLYOLEFINS

	Low density polyethylene	High density polyethylene		Polypropylene
		Phillips	Ziegler	
Density	0·91–0·93	0·96–0·97	*ca.* 0·95	0·90–0·91
Tensile strength (lb/in²)	1000–2300	3100–5500	4500	4200–5500
Impact strength (ft-lb/in² of notch at 15°C)	No break	5	3	0·4–7
Dielectric strength (V/mil)	460–700	>500		>800
Dielectric constant (10⁶ cycles/sec)	2·25–2·35	2·25–2·35		2·25–2·3
Power factor (10⁶ cycles/sec)	<0·0005	<0·0003		<0·0002–0·0003
Volume resistivity (ohm/cm)	10¹⁵	10¹⁵		10¹⁵
Softening point (°C (Vicat))	85–87	120–130		150
Melting point (°C)	110	130		170
Heat distortion temperature (°C)	40–49	60–82		99–112

point. In the case of polypropylene the increase in low temperature impact strength is substantial. In the case of high density polyethylene, while the impact strength of high molecular weight homopolymer is satisfactory, a lower molecular weight copolymer may be used with consequent improvement in processing properties.

Figures have not been given for elongation at break since this property can vary widely among different grades of the same polymer. In general, low density polyethylene may have values around 400% while the other polymers give figures substantially lower than this. In this connection it may be noted that all the materials show some tendency to creep under static load and this is particularly true of the more linear polymers, especially when the molecules have been deliberately orientated by stretching. It may be controlled by introduction of some chain branching which restricts the movement of the molecules relative to each other.

All the materials have excellent electrical properties, high density polyethylene and isotactic polypropylene being slightly superior to low density polyethylene.

The differences in softening point between the polymers have a major effect on their fields of application. All three have glass transition temperatures below room temperature (both polyethylenes $-85°C$ and polypropylene $0°C$). Low density polyethylene cannot be used above 70–75°C, whereas the high density material has a softening point just high enough to permit objects made from it to be sterilized; polypropylene may be used continuously in boiling water.

Chemical Resistance

Low density polyethylene is resistant to all inorganic chemicals except chlorine and concentrated nitric acid at room temperature but absorbs most hydrocarbons and their halogen derivatives with consequent swelling. Above about 70°C it dissolves in most organic solvents, forming gels on cooling.

Chemical resistance increases with both crystallinity and molecular weight. High density polyethylene does not dissolve in solvents, even when heated, and swells very little at room temperature. Polypropylene is rather less resistant and resembles low density polyethylene in this respect. Molecular weight plays a very important part in determining chemical resistance and rather outweighs the difference between the polymers. Very high molecular weight low density polyethylene swells very little.

A small amount of swelling does not necessarily render the polymer unsuitable for carrying the liquid concerned—in a pipe for example; what is more serious is the surface cracking which sometimes occurs when the polymers are stressed while in contact with some liquids and vapours. In this respect materials of very high molecular weight do show to considerable advantage. The trouble may be overcome by using a high density copolymer of ethylene with a little propylene.

Other Properties

All polyolefins burn slowly in air, polypropylene being the fastest burning, and all of them are affected by sunlight except when compounded with carbon black; for transparent materials exposed to sunlight light stabilizers must be used. High density polyethylene and polypropylene have excellent machining qualities while both the low density polymer and polypropylene are better than high density polyethylene for moulding.

APPLICATIONS OF POLYOLEFINS

There is much overlapping in the fields of application of the three polymers and the choice of one or the other usually depends on the cost–performance characteristics required for the job. Normally, low density polyethylene is the cheapest and polypropylene the most expensive material expressed as the price per pound. However, the different grades of each of the polymers vary considerably in price so that each case has to be treated individually.

Film and Sheet

More than half the U.K. consumption of low density polyethylene goes into this application—about 87,000 tons in 1967. Very little high density polyethylene is used in this way but it is an important outlet for polypropylene and absorbed some 5000–6000 tons, or rather over 10% of U.K. production, in 1967. The distinction between film and sheet is an arbitrary one—below a thickness of $0 \cdot 01$ in. it is film, above this thickness it is termed sheet.

Low density polyethylene film can be made either as "lay flat" tubing or by a cast film process. In the former the polymer is extruded vertically upwards through a circular die; the tube is drawn off through a pinching device some distance from the die and is eventually wound on to a reel as a flattened tube. Air is blown into the tube through the centre of the die and is trapped by the pinching device causing the tube to expand; by regulating the speed of extrusion and the air pressure any desired tube diameter and wall thickness can be produced. The lay-flat tubing may be sold as such or it may be slit to form a single layer sheet or film. The process is illustrated in Fig. 22.

In the cast film process molten polyethylene is extruded downwards through a narrow slit on to a water cooled roll, where it solidifies, and is usually carried over one or more additional cooled rolls to complete the cooling. Film made in this way is clearer and glossier than the blown film.

About 20% of the film produced is used for the production of sacks, either by itself or as a lining for sacks made of paper or other material, where it provides resistance to water. A large part of the remaining production goes into general packaging, its freedom from odour and resistance to fats and oils making it particularly suitable for prepacked foodstuffs. Agriculture, too, absorbs substantial quantities of film where it is used as seed-bed covers, for conserving mulches, as a waterproof lining for ponds and ditches and in various arrangements for the temporary storage of crops.

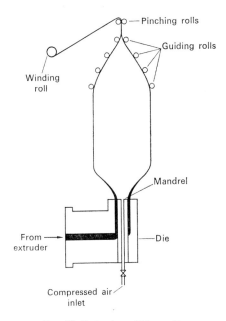

FIG. 22. Extrusion of blown film.

Another major outlet is in the building industry where it is used for damp proof courses under concrete floors, for covering newly laid concrete to promote even curing and for various roofing applications. It can also be used to enclose scaffolding so that building work may proceed in bad weather. Many instances of its use to protect cars left in the open can be seen in our crowded and garageless suburbs.

Polyethylene for film manufacture is usually of medium molecular weight with a melt index (see p. 55) between 0·6 and 5.

Polypropylene does not form a satisfactory lay-flat tubing when the method described above for low density polyethylene is used. The slow cooling produces a hazy product with a low tear resistance. Special equipment has now been developed in which the cooling is much more rapid and on which polypropylene lay-flat tubing can be produced satisfactorily. Polypropylene film of excellent clarity and toughness can also be produced by the casting technique. The films are mainly used for packaging, especially for foodstuffs.

As noted on p. 166, the strength of polypropylene film can be greatly increased by stretching or drawing. Biorientated film is made by first stretching the film in the machine direction, usually by passing it over pairs of rolls rotating at different speeds, the film being heated between one pair of rolls and the next. The drawn film shows a reduction in width and it is now stretched in the transverse direction on a divergent tenter frame while being prevented from shrinking in the machine direction. Although highly crystalline, the films are very transparent since the orientation of the molecular chains brought about by the stretching prevents the crystallites arranging themselves into aggregates.

Biorientation is an expensive operation and, for most purposes, ordinary film is quite adequate. The additional strength of biorientated film (tensile strength up to $24,000 \, lb/in^2$) does, however, permit a thinner film to be used while its high crystallinity reduces permeability to gases, oil and grease as well as improving the impact strength. A major disadvantage is that it cannot be heat sealed since it shrinks as it melts; this can be overcome by coating it with a heat sealable polymer with a melting point below that of polypropylene, but at increased cost. Biorientated film is competitive with regenerated cellulose film (which also has to be coated for heat sealing). An interesting, though small, use is in the electrical industry where very thin gauge film is used as the dielectric in condenser manufacture.

Injection Moulding

This is a major outlet for all three polymers and consumed over 30,000 tons of low density polyethylene, 12,000 tons of high density polyethylene and between 12,000 and 15,000 tons of polypropylene in 1967. A polymer of rather higher melt index (lower molecular weight) is required than for film manufacture but a narrow molecular weight range is preferred; this aids moulding speed and makes for more even cooling and shrinkage. Standard injection moulding techniques are used, a slightly higher temperature being required for high density polyethylene and polypropylene than for low density polyethylene.

The method is used for the production of a wide range of industrial articles such as buckets, basins and containers of all shapes up to the maximum size of injection moulding machine available. Some high frequency insulators are moulded; these can be of any polyolefin although polypropylene is probably slightly superior and has, in addition, a higher heat resistance.

Low density polyethylene had been used for injection moulding for some years before the other polyolefins became available. The extent to which it is now being replaced by the other polymers depends mainly on economics and, in particular, on the value which can be placed on their greater rigidity and resistance to heat. Now that the development of copolymers has overcome the earlier disadvantages of high density polyethylene and polypropylene, they are attractive for injection moulding applications although, for the man in the street, it is not easy to distinguish between them when seen, for example, in the form of a washing-up bowl; both can be used in contact with boiling water and they are of approximately equal stiffness.

Blow Moulding

Eighteen thousand tons per annum each of low and high density polyethylenes and 6000 t/a of polypropylene go into this outlet. The technique is used mainly for the manufacture of bottles and similarly shaped containers with a relatively large internal capacity and a narrow opening. Here the resistance of high density polyethylene to both sterilization and low temperatures shows to particular advantage.

Extrusion

Orthodox extrusion techniques are mainly confined to the production of continuous rods, tubes and special profile shapes and for the formation of monofilament and fibres. High molecular weight polymers with a wide molecular weight distribution give the best compromise, the proportion of relatively low molecular weight material present aiding flow. Low density polyethylene is not normally extruded into sheet but excellent sheet may be produced from polypropylene by standard procedures.

Cable covering is a special kind of tube extrusion, a specially designed extruder in which the wire is fed through the centre of the die being used. About 17,000 t/a of low density and 2000 t/a of high density polyethylenes are used for this purpose. Other tubes and pipes take about 5000 t/a of low density and 2000 t/a of high density polyethylenes but only a very small quantity of polypropylene. The pipes are used for carrying water and many corrosive chemicals at normal, or very slightly elevated temperatures and mainly at moderate pressures, although under the proper conditions high density polyethylene pipes can be used for liquids at substantial pressures. Polypropylene shows promise of being usable at higher temperatures than the other polymers but its brittleness at low temperatures has been conducive to caution. Much work is going on to develop polypropylene for carrying water up to its boiling point but trial installations have not yet been in use for long enough and it will be some time before a clear picture emerges. If the trials are successful its potential for domestic hot water systems is obviously considerable.

Some 8,000–9000 t/a of low density polyethylene are used for the coating of paper, aluminium foil and similar materials. The polymer is extruded directly as a thin coating on to the surface of a moving band of the material to be coated in equipment rather similar to that used for making cast film. The coated band passes immediately through rolls while still warm to seal the coating to the substrate. These materials are widely used for milk cartons, food sachets and similar applications.

MODIFIED POLYETHYLENES

Many attempts have been made, especially with low density polyethylene, to produce modified polymers with the object of improving some particular property such as softening point, resistance to solvents and sunlight and so extending its use. One of the most interesting is the subjection of the ready formed polymer as sheet, pipe or other convenient shape, to high energy radiation. This causes a certain amount of cross-linking of the polymer chains, which can be controlled by controlling the radiation dose, and increases both the melting point and solvent resistance; if carried far enough the process produces an infusible and insoluble material like a thermosetting resin.

Chemical modification of the polymer has been attempted by such processes as oxidation, halogenation and sulphonation and the patent literature contains many claims for the production of useful materials. One of these, at least, a chlorosulphonated low density polyethylene, has been marketed by DuPont in the United States under the trade name of "Hypalon".

In general it may be said that none of the modified materials has yet achieved really large scale commercial acceptance.

COPOLYMERS

The production of copolymers of ethylene with a little propylene and of propylene with a little ethylene by the low pressure processes has been described earlier in the chapter. These copolymers have been a great success commercially and their use is likely to go on growing rapidly for some time; they must not be confused with the elastomeric copolymers, described in Chapter 13, in which the proportions of ethylene and propylene are more nearly equal.

Copolymers of ethylene with vinyl acetate have achieved some success in the United States, and I.C.I. are just starting production in the United Kingdom. They are described more fully on p. 223.

OTHER POLYOLEFINS

Although most olefins can be made to polymerize, none of them has so far seemed likely to approach ethylene and propylene in importance. A number of polymers have, however, achieved some commercial success and are briefly described.

Poly 4-Methyl Pentene

The manufacture of the monomer has been mentioned on p. 150. This is a new material which appears to be on the point of reaching commercial production. It is the lightest of all commercial thermoplastics with a density of $0 \cdot 83$; the melting point is higher than that of polypropylene (205°C against 170°C) but it is more brittle at normal temperatures. Transparency is almost as good as that of glass and this gives the polymer great potential. I.C.I. have erected a moderate scale plant to make 2000 t/a at Wilton, Yorks.

Polybutenes

There are three groups of polybutenes of which two have achieved some commercial development and the third is potentially interesting.

Low Molecular Weight Polybutenes

A mixture of C_4 olefins, often containing some C_3 and C_5 olefins, is polymerized cold with acid aluminium chloride. By varying the temperature, molecular weights between 300 and 1500 can be obtained. The polymers are viscous liquids, mainly used as viscosity index improvers in lubricating oils. They are made by B.P. Chemicals (U.K.) Ltd. at Baglan Bay, Glamorganshire.

Polyisobutylene

Isobutene, commonly called isobutylene in industry, can be separated from other C_4 olefins by its greater solubility in 60%

sulphuric acid at normal temperatures. The acid solution forms a separate layer which may be run off and from which di-iso-butylene may be recovered on heating; the dimer may be re-converted to the monomer by mild pyrolysis.

Most isobutylene is used to make butyl rubber by copolymeriza-tion with 4–6% isoprene but it may also be polymerized alone at −80°C in the presence of boron trifluoride as catalyst; liquid ethylene is used as solvent and as a means of controlling the temperature. This reaction is a good example of cationic poly-merization. Rubbery polymers are formed which are flexible down to −6°C and are resistant to oxygen and ozone. On the other hand they show considerable cold flow, have limited compatibility with other polymers and cannot be vulcanized owing to the absence of any residual unsaturation. Their main applications are in pressure sensitive adhesives, paper coatings and caulking compounds. They are not made in the United Kingdom.

Poly 1-Butene

Using Ziegler-Natta catalysts l-butene can be polymerized to an isotactic, highly crystalline polymer which has a density of 0·89–0·91 and softens and melts at about the same temperature as Ziegler polyethylene. It has an elongation at break of about 300%, thus showing some elastomeric properties unexpected in such a highly crystalline polymer. Electrical properties and chemical resistance are similar to those of polypropylene. The polymer has not been developed commercially, possibly because of the cost of isolating l-butene in a sufficiently pure state.

POLYOLEFINS IN THE FUTURE

Reviewing the current state of development of polyolefins it appears likely that there is still room for appreciable further exploitation of the three materials which have already achieved major commercial importance. Low density polyethylene has already taken its place, with polystyrene and PVC, as one of the three leading thermoplastics. Its range of properties and its applications are still being extended so that a relatively high rate of growth in consumption is likely to continue for some years yet.

High density polyethylene and polypropylene, although over-shadowed by their bigger brother, have achieved a remarkable rate of growth in the limited period that they have been available in commercial quantities. The possibilities of technical improvement of the polymers have by no means been exhausted and the process of cost reduction has not yet been carried so far as with the low density material. For these polymers, too, a period of high growth rate extending over a number of years can be predicted.

The possibility that other polyolefins will be developed to rival the polyethylenes and polypropylene seems, on the face of it, unlikely. For one thing, the cost of more complex olefin monomers is always likely to be substantially higher than that of the two lower members of the series. The somewhat unexpected properties of the polymers of 1-butene and 4-methylpentene show, however, that substantial commercial development of other olefin polymers cannot be entirely ruled out.

READING LIST

The Discovery of Polythene, by R. O. Gibson, Royal Institute of Chemistry Lecture Series, 1964, No. 1.
Polythene, by Renfrew and Morgan (2nd edn.), Iliffe Press, London, 1960.
Polypropylene, by Kreiser, Reinhold, New York, 1960.
Crystalline Chain Polymers, by Raff and Doak, Interscience, New York, 1964.
The Stereochemistry of Macromolecules, by Ketley, Arnold, London, 1967.

CHAPTER 9

Styrene Polymers and Copolymers

POLYSTYRENE was actually produced on a laboratory scale as early as 1845 but both the polymer and monomer remained curiosities until the second quarter of the twentieth century. Large-scale production of styrene did not develop until the late twenties and much of the early polystyrene had poor ageing properties. Although the pure monomer autopolymerizes very easily unless inhibited, the production of a satisfactory commercial polymer is much more difficult than might be supposed and really acceptable polymer was not made until the mid-thirties, first by B.A.S.F. in Germany and then by Dow and Bakelite in the United States. In the United Kingdom Standard Telephones and Cables Ltd. had been jointing high frequency cables during the thirties by polymerizing styrene, mainly supplied by A. Boake Roberts and Co. Ltd. and made by the dehydration of β-phenyl ethanol.

D.C.L. started pilot plant production of polystyrene by bulk polymerization at Tonbridge, Kent in 1938 at the rate of 50 t/a. After some increase in capacity, production was maintained throughout the war at the rate of 150–200 t/a according to the availability of monomer, most of which was imported. This polystyrene was used for radar accessories, range-finder prisms, instrument dials, eye shields for divers' helmets and other applications connected with the war effort. The plant was transferred to Barry in 1948–9 and its capacity increased to about 1000 t/a. An emulsion process, giving a high molecular weight product, was also operated on the same site. In 1953 D.C.L. joined forces with Dow, Distrene Ltd. was set up and the

Dow processes was operated at Barry until 1967–8 when BP acquired most of the D.C.L. chemical interests. The Distrene plant at Barry, however, became 100% Dow owned and is continuing to operate.

Soon after the Second World War both Monsanto and, a little later, Styrene Products Ltd. began to manufacture polystyrene; the latter was a joint company owned by Erinoid Ltd. and Petrochemicals Ltd. In 1955 the Shell Group acquired Petrochemicals and, later, the Erinoid interest in Styrene Products so that the latter became part of Shell Chemicals (U.K.) Ltd. Erinoid, after various vicissitudes, became part of B.P. Chemicals (U.K.) Ltd. and has its own plant at Stroud, Gloucestershire. Sterling Moulded Materials Ltd., who entered the field rather later with a plant at Stalybridge, Cheshire, has recently taken over the polystyrene activities of the B.X.L. Group at Manningtree, Essex. The site at Manningtree, used by British Xylonite Ltd., later B.X.L. Plastics Materials Ltd., for the production of nitrocellulose and celluloid, began producing polystyrene in 1953. The company eventually became a 100% subsidiary of D.C.L. until the creation of the Bakelite Xylonite complex in 1962 (see p. 60). The acquisition of these polystyrene operations by Sterling Moulded Materials Ltd. is also part of the disposal of the D.C.L. chemical and plastics interests referred to above. The present manufacturing capacities of the companies involved are believed to be as given in Table 18.

TABLE 18. U.K. MANUFACTURING CAPACITY FOR POLYSTYRENE
(figures in 000 t/a)

Dow (U.K.) Ltd.	Barry, Glam.	40
Shell Chemicals (U.K.) Ltd.	Carrington, Cheshire	36
Monsanto Chemicals Ltd.	Newport, Mon.	35
Sterling Moulded Materials Ltd.	Stalybridge and Manningtree	34
B.P. Chemicals (U.K.) Ltd.	Stroud, Glos.	18
Kaylis Ltd.	Bolton, Lancs	6
Total capacity		169

THE MONOMER

Several methods of synthesizing styrene are known but, for commercial production, the method of alkylating benzene with ethylene to form ethyl benzene followed by catalytic dehydrogenation is almost universally used.

Catalytic dehydrogenation is carried out in the presence of steam at about 600°C. The styrene monomer is purified by vacuum distillation, as it polymerizes quite rapidly at 145°C, its boiling point under normal pressure; the distillate is immediately inhibited, usually with about 10 parts per million of *p*-tertiary butylcatechol, before being sent to storage. The inhibitor must be removed by a preliminary distillation before polymerization. There are at present only two manufacturers of styrene in the United Kingdom—Forth Chemicals Ltd. (owned two-thirds by B.P. and one-third by Monsanto) and Shell Chemicals (U.K.) Ltd. The former has two plants, at Grangemouth in Scotland and Baglan Bay in South Wales, with capacities of 60,000 and 75,000 t/a respectively and the Shell Chemicals plant at Carrington, Cheshire, has a capacity of 76,000 t/a. The International Synthetic Rubber Co. Ltd. has, however, announced plans for a plant with a capacity of 50–60,000 t/a.

POLYMERIZATION

Styrene polymerizes very readily under the influence of heat alone, or a wide range of catalysts may be used for polymerization at lower temperatures. Four processes, described as bulk, suspension, emulsion and solvent methods, may be used, of which the first two are the most popular. Molecular weights from a few thousands to well over 2 million may be obtained by adjusting the conditions of polymerization; other things being equal, the higher the temperature of polymerization the lower the molecular weight. Thus polymerization by heat alone has been found to give an average molecular weight of 130,000 when carried out at 140°C whereas, at 60°C, values as high as 2,250,000 are obtained; the polymerization rate in the former case is 300 times that of the latter. The four methods are briefly described in the following paragraphs.

Bulk Polymerization

This is generally carried out by heat alone, either batchwise or in a continuous process. The polymerization reaction is exothermic, and the difficulties of heat removal from material which becomes more and more viscous as the reaction proceeds means that batches must be small. Even so there is a relatively large mass of polymer to be broken up in some way when the polymerization is complete. In the continuous process the monomer is first partially polymerized in batches at 80–90°C to give a viscous solution of polystyrene in unpolymerized monomer; this is then passed continuously down a tower in which the temperature increases at a controlled rate as it passes downwards and the polymerization is completed. Polymer is discharged from the bottom of the tower at about 200°C, partially cooled and forced through dies to form strands which can be chopped into small pieces. The extrusion is illustrated in Plate IX. Either method gives a product of high clarity and excellent electrical properties but with a fairly broad distribution of molecular weight. Care must be taken to remove the last trace of monomer to produce a stable product.

Suspension Process

This is carried out in simple kettles fitted with stirrers; the monomer is broken up into droplets suspended in water with a styrene-soluble catalyst, such as benzoyl peroxide, and water-soluble polymers and similar materials to assist in stabilizing the suspension. Each styrene droplet has some catalyst in solution

PLATE IX. Extrusion of polystyrene into strands. The strands can be seen issuing from the extruder head into the cold-water bath. (By courtesy of Shell Chemicals Ltd.)

and is, in effect, a very small scale bulk polymerization reactor while the presence of a relatively large volume of water makes temperature control much easier. The process produces small beads of polystyrene which may be used directly for injection moulding; usually, however, they are required in a range of colours and are compounded with dyestuffs and pigments.

Emulsion Polymerization

This is somewhat similar to the suspension process but the monomer is broken up into much smaller droplets to produce a true emulsion in water with the aid of an emulsifying agent; a water-soluble catalyst is used and buffering agents are also added as the process is affected by changes in the pH value. The product is inferior in clarity and colour to that from the processes already described; its molecular weight is high and may be too high for some purposes so that special chain stoppers have to be added during polymerization. The polymer is produced in the form of a very fine powder which it is difficult to free from contamination with the various reagents used during the reaction. A slight advantage of the process is that the product has a somewhat higher heat distortion temperature than normal. The process is also a useful way of making copolymers; its use is largely confined to this operation and to those few cases where a polystyrene emulsion is required as the finished product.

Solvent Process

The styrene is polymerized in solution in an inert solvent such as cyclohexane; the solutions become very viscous as the reaction proceeds but temperature control is easier than with the bulk process; while the product tends to have a low molecular weight, the nature of the solvent has a very important effect. Removal of the last 10% of solvent from the polymer is not always easy. The process is used mainly in those applications of coating and impregnation where a solvent would be required in any case.

THE POLYMER

Polystyrene is a colourless and glass-clear material showing signs of distortion at 75–80°C and softening at about 100°C; it has a high tensile strength, around 6000–8000 lb/in², but its impact strength is low. The polymer is very resistant to inorganic chemicals, even to the strongest acids, but is affected by most organic solvents; it is actually soluble in aromatic and chlorinated hydrocarbons. It is very resistant to water, has excellent dimensional stability and outstanding electrical properties, its volume and surface conductivity and power factor being virtually zero. It can also be dyed to give most attractive shades without affecting the transparency but a light stabilizer is required if the product is to withstand long outdoor exposure.

General Applications

The moderate softening temperature of the polymer, and the fact that it softens over a considerable temperature range, make it an almost ideal material for injection moulding, especially for small articles and, in the decade after the war, it virtually displaced cellulose acetate in this field. Its high refractive index (1·6) and transparency make it particularly useful in the production of ornamental articles; it has been used for light fittings, reflectors, novelties of all kinds, toys and so on. Much polystyrene is extruded as sheet and this can be vacuum formed to produce fancy boxes and for many packaging applications. In its early days, because it was freely available and relatively cheap, it was used in many applications for which it was not suitable and this brought disrepute, not only on polystyrene, but on all plastics, in the immediate post-war years. Straight polystyrene as described above has two major disadvantages:

(i) Its low impact strength combined with the sharp edges produced when it is fractured.
(ii) Its relatively low softening point which makes it unsuitable in some electrical applications for which its other properties would have made it the ideal material.

Much work has gone into improvement of the material which can be dealt with under three broad headings: modification of the polymer itself, modification by compounding the polymer with other materials or by production of copolymers and polymerization of analogues and homologues of styrene. Direct improvement of the polymer itself clearly has limited possibilities but efforts to achieve it have led to the production of foamed polystyrene which has been outstandingly successful and has opened up an entirely new range of uses for the polymer.

Foamed Polystyrene

Polystyrene foam is produced by incorporating a volatile liquid into the solid polymer to make what is known as an expandable grade; when this is heated the included liquid vaporizes as the polymer softens and small bubbles of vapour expand the whole mass into a rigid, lightweight, closed cell foam. The process can be divided into two definite stages: formation of the expandable material and expansion into the desired form. The expander, or "blowing agent", is usually a low boiling liquid such as pentane or methyl chloride and is added to the extent of 5–8% on the polymer. It may be incorporated by allowing it to diffuse into beads of polystyrene from the suspension process or it may be added to the monomer which is then subjected to a suspension polymerization. The expandable beads so formed may be placed directly in a mould and heated when they will expand to the required shape. This method, however, does not give a product of uniform density and better results are obtained by pre-expanding the beads before moulding. Pre-expansion may be carried out continuously by feeding beads into the top of an upright drum where they meet a rising current of steam which causes them to soften and expand. An agitator keeps the beads free flowing and prevents them fusing together while pre-expanded beads are discharged continuously from the bottom of the drum. When a mould is filled with these roughly spherical beads, closed and heated under a pressure of about 20 lb/in^2, no further overall expansion is possible, but expansion of the vapour contained in the bubbles and evaporation of residual blowing agent will cause the beads to expand to fill the voids between the spheres and, in their softened condition, they will fuse together to form a coherent mass. Expandable beads may be extruded through a heated rectangular chamber to form a continuous block which can be cut into convenient lengths and sliced to form sheets of any desired thickness. Another method of forming sheet which is being developed is to feed polystyrene beads and blowing agent together into an extruder slit so that a continuous expanded sheet is formed; this method is particularly applicable to thin sheets.

PROPERTIES OF THE FOAMED MATERIAL

The physical properties of the expanded product depend to a considerable extent on the amount of blowing which has taken place. The density may vary from 1 to 6 lb/ft^3 and the strength increases almost linearly with the density; at 4 lb/ft^3 the compressive strength is 60 lb/in^2, tensile strength 100 lb/in^2 and flexural strength 125 lb/in^2. It is a closed cell foam, so that it does not absorb water like a sponge which has open cells; it has a useful strength/weight ratio and a very low thermal conductivity of about $0 \cdot 25$ B.t.u./ft^2/°F/in.; it retains its properties down to about -70°C.

As might be expected, the foam has very poor heat resistance and it will soften and collapse at quite moderate temperatures; it will also burn quite freely. Flammability can be reduced by incorporating certain bromine compounds to produce a self-extinguishing grade—that is to say a material which will not continue to burn once the source of ignition is removed—but this will not, of course, prevent its softening and collapse in a fire.

APPLICATIONS

Foamed polystyrene is, to some extent, competitive with the more recently developed rigid polyurethane foams which are described in Chapter 7. Generally, foamed polystyrene is better for thermal insulation and has greater resistance to water vapour penetration but is inferior for heat resistance and flammability. It is used in a wide range of applications of which only a few are described below.

Thermal Insulation

Foamed polystyrene is widely used for the insulation of fixed cold stores, refrigerated road and rail vehicles, air conditioning equipment and in many situations where the temperature is below, or only a little above, the normal atmospheric level.

Buoyancy

It can be used, suitably covered to protect it from abrasion, to provide buoyancy in small boats, life saving equipment and swimming pool accessories and for the manufacture of floating toys. It has recently been used in refloating sunken vessels.

Packaging and Containers

Lightness associated with strength and softness combined with rigidity make foamed polystyrene an ideal material for a wide range of packaging application. Blocks of the material may be moulded precisely to the shape of delicate articles and thus give a degree of protection against shock which can never be attained with the more conventional packing materials. Loose expanded beads are also excellent for packing irregularly shaped articles where the more expensive moulded form is not necessary.

Although the material is quite rigid in block form, thin sheets may be made of varying degrees of flexibility. In this latter form it may be used for packing eggs, in the production of carton inserts for packing such things as chocolates and biscuits and for the manufacture of excellent disposable cups and jars. It may also be used as a core between paper stock to form a sheet from which containers may be produced.

Building

Foamed polystyrene has many uses in the building industry, most of which depend on its good thermal and sound insulating properties. Compressed thin sheets are applied to interior walls, rather like wall paper, where they are excellent for preventing heat loss and condensation; decorative finishes are applied over the polystyrene foam. Thicker sheets may be used for ceiling insulation and, sandwiched between sheets of plywood or similar material, for internal partitions or, suitably protected, for outdoor applications.

POLYSTYRENE LATEX

This product, which is quite different from the copolymer latices mentioned on p. 193, can be made by emulsion polymerization of styrene. It is manufactured commercially in limited quantities and has been used in glass-reinforced products, polishes and in certain kinds of surface coatings.

LOW MOLECULAR WEIGHT POLYSTYRENE

Polystyrene of molecular weight 50,000–100,000 (compared with 200,000–300,000 for the normal quality) is used in lacquers, for coating paper and, mixed with asphalt, for the manufacture of floor tiles.

MODIFIED POLYSTYRENE AND COPOLYMERS

The two properties of polystyrene which most require improvement are its impact strength and softening point. The former can be substantially improved either by compounding or by copolymerization but only limited improvements in softening point and heat distortion temperature have been possible by either method. Compounded materials, collectively classified as "high impact" polystyrenes, contain relatively small amounts of toughening agent but it is not always clear whether the copolymers should be regarded as modified polystyrenes or as separate materials in their own right. For statistical purposes the Board of Trade, advised by the British Plastics Federation, treat any polymer containing 50% or more of styrene as "Polystyrene"; this is not, however, entirely satisfactory as some rubber-like copolymers of styrene and butadiene (SBR) and the terpolymers of acrylonitrile, butadiene and styrene may exceed this limit for styrene content. The butadiene–styrene copolymer containing 25% of styrene is, of course, the well-known general-purpose synthetic rubber and is described, with other synthetic elastomers, in Chapter 13. The acrylonitrile–butadiene–styrene terpolymers, generally known as ABS plastics, have become sufficiently important to be treated as plastics in their own right and are dealt with in a separate section at the end of this chapter.

High Impact Polystyrene

It might be thought that the incorporation of some of the well-known plasticizers with polystyrene would give a product with higher impact strength, but the only materials compatible with the polymer have the effect of lowering the softening point and impairing the hardness of the material. The incorporation of synthetic rubber has, however, proved to be a good solution to the problem of obtaining a material of higher impact strength. The synthetic rubber chosen is usually the ordinary SBR general purpose synthetic rubber, referred to in the preceding paragraph, which is made on a scale exceeding 100,000 t/a in the United Kingdom, largely for tyre manufacture. As an alternative, poly-butadiene is coming into use for certain goods (see p. 275).

Two methods of incorporation are used; the synthetic rubber can be blended with the polystyrene by fluxing on mixing rolls or it may be dissolved in the styrene monomer before polymerization. The former method gives some increase in impact strength while the latter technique can give a two- or three-fold increase. Both methods, however, give products which are no longer transparent and this reduces their value for some applications;

the processing characteristics are not otherwise seriously impaired.

In the second method of incorporation, the synthetic rubber is broken up into very small pieces and is dispersed in the styrene monomer with the addition of a small quantity of toluene; 10–20% is a typical figure for the proportion of rubber added. A solvent polymerization is carried out and the resulting viscous liquid is transferred to a batch still, where it is heated under moderate reflux to polymerize the last traces of the monomer; the solvent is then distilled off. The semi-liquid high impact polystyrene is extruded into strands, chopped and the resulting nibs worked up in the same way as ordinary crystal polystyrene.

PROPERTIES AND USES

High impact polystyrene from the solvent polymerization method may have an impact strength of $0 \cdot 4$–$3 \cdot 0$ ft-lb/in. of notch compared with a figure of $0 \cdot 2$–$0 \cdot 5$ for the ordinary crystal material. Due to its excellent resistance to staining and to attack by food products, it is used for refrigerator linings and for disposable cups, food containers, bottle caps and many other similar uses in the food industries; although no longer transparent the material has a pleasing gloss, good fatigue strength and may be produced in pale colours. These properties, combined with good electrical resistance, make it attractive for the cabinets housing air conditioning equipment, portable TV and radio sets and for the grills, dials, control knobs and similar component parts of such sets. More generally it may be used for wall tiles, in the construction of storage batteries, shoe heels, furniture legs, many kinds of houseware and toys and the cheaper types of gramophone records; monofilaments have been extruded and used for brushes.

High impact polystyrene is a relatively cheap and versatile thermoplastic which has greatly extended the range of application of polystyrene. Many other uses could be added to the selection made above which will, it is hoped, give some idea of the scope for the material.

POLYMERS WITH IMPROVED HEAT RESISTANCE

Attempts to improve the heat resistance of polystyrene have not had the same success as those to raise the impact strength and no commercially viable simple solution has been found. Polymerization of dichlorostyrene gave promising results but was too expensive for other than highly specialized applications. Another approach was copolymerization of styrene and *p*-divinyl benzene; even 1% of the divinyl compound raised the heat distortion temperature to 115°C but thermoplasticity was greatly reduced since the additive acted by cross-linking the polymer chains.

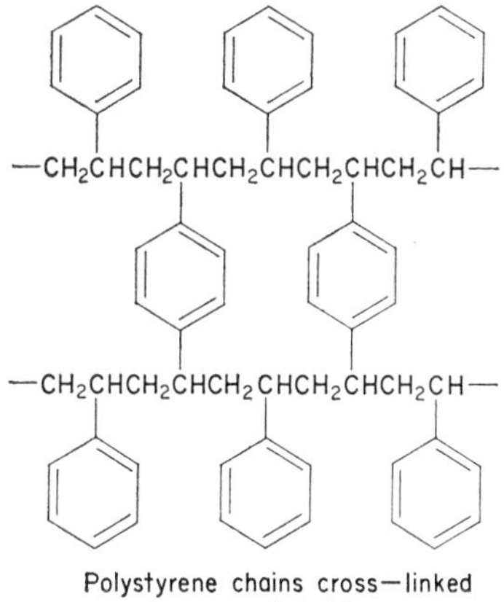

Polystyrene chains cross—linked
by para—divinyl benzene

This made processing difficult and outweighed any advantage gained from extra heat resistance.

Styrene–acrylonitrile copolymers, containing about 10% of acrylonitrile, made by suspension or bulk polymerization techniques, have achieved limited success. The heat distortion temperature is increased to about 95°C and the copolymers are more resistant to hydrocarbon solvents than polystyrene. They are, however, slightly more difficult to process and slightly less water resistant; if the acrylonitrile content exceeds 20% the colour stability at higher temperatures is also poorer. United Kingdom production amounts to a few hundred tons per annum. One of the outlets is for manufacture of picnic ware.

HOMOLOGUES OF STYRENE

The homologues of styrene have been examined to see if their polymers offer any advantages. The two simplest are α-methyl styrene and vinyl toluene; in the latter the methyl group is attached to the benzene nucleus. α-Methyl styrene may be made from propylene and benzene in the same way as styrene from ethylene and benzene.

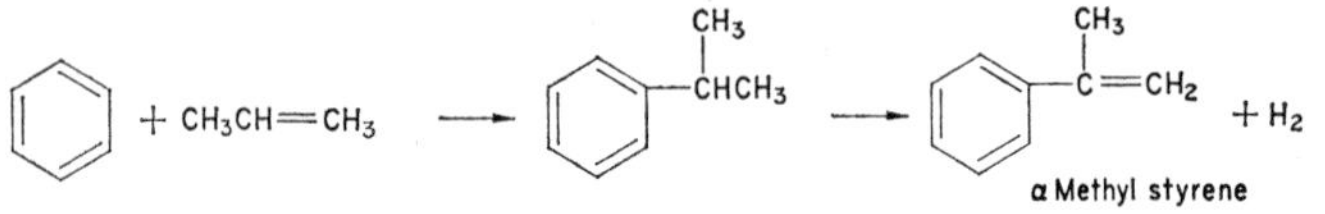

α Methyl styrene

It is a liquid, boiling at 165°C, and should be available cheaply if large scale manufacture were developed, but, unfortunately, it offers no advantages. It does not form high polymers by free radical mechanisms, but polymers of a very high molecular weight may be produced by cationic polymerization with a Friedel Crafts type catalyst in ethyl chloride at $-70°C$; they have much higher softening points but are exceedingly difficult to process. Copolymers of styrene with up to 30% of α-methyl styrene soften above 100°C; they may be made by emulsion polymerization but their flow properties are impaired and they are not easy to process. For this reason they are not popular.

The three isomers of vinyl toluene differ in the properties of their polymers; the ortho-isomer forms a polymer with markedly higher heat distortion temperature than polystyrene and, if the monomer were available at a suitable price, might achieve commercial importance. Unfortunately, production of vinyl toluene from toluene and ethylene by a process analogous to that for styrene yields about 65% of the meta-isomer, 35% of para and negligible ortho. This mixture is useless for the production of polymers of improved heat distortion temperatures but is valuable for incorporation in unsaturated polyesters and styrenated alkyds (see pp. 101 and 99 respectively). A large scale process was operated in the United States for a short time in which a mixture of *o*- and *p*-vinyl toluene was produced by reaction of toluene and acetylene followed by vapour phase cracking. Formation of the orthoisomer may be represented as follows:

CH_3 + $CH \equiv CH$ ⟶ CH_3 CH_3 ⟶ CH_3 $-CH = CH_2$ + CH_3

o-Vinyl toluene

The process was abandoned on the grounds of cost and, possibly, poor yields.

STYRENE BUTADIENE COPOLYMERS

Styrene and butadiene may be copolymerized in all proportions; emulsion polymerization produces latices which may be used for a variety of purposes. Up to about 55% styrene, the copolymers show rubbery characteristics and are dealt with under "Elastomers" in Chapter 13. Between 55% and 70% styrene, however, the latices have film-forming properties and have been widely used in the United States as a basis for emulsion paints; this use was stimulated by the ready availability of styrene monomer from the huge U.S. synthetic rubber programme. These emulsions have made little progress in the United Kingdom as vinyl acetate emulsions are normally considered superior. They are, however, used for paper coating as they give a smoother printing surface, better folding characteristics and reduced curl when compared with other paper-coating emulsions.

Copolymer emulsions with still higher styrene content have been used to some extent, both for paper coating and for reinforcing rubber. The reinforced rubber products were tried out extensively for shoe soles but the adhesion of the sole was disappointing and it is believed that these materials have now been virtually abandoned.

ION EXCHANGE RESINS

This is a small-tonnage but very interesting application for styrene/divinyl benzene copolymers. A cross-linked copolymer, as described on p. 191, is first formed and this is then sulphonated. The resulting resin has the property of removing cations from aqueous solution, replacing them with hydrogen ions. When the resin becomes exhausted, its ion exchange power can be re-generated by treatment with an acid.

If, instead of sulphonation, the hydrocarbon resin is first nitrated and then reduced to give an amino substituted material, anions may be removed from solution and replaced by OH groups. The net effect of consecutive filtrations through these two materials is a demineralized water. A simple modification is to regenerate a cationic exchange resin with sodium chloride when it will replace calcium with sodium ions, thus softening the water.

FUTURE FOR POLYSTYRENE

The U.K. production and consumption of polystyrene over the past 14 years is shown in Table 19.

The figures include expanded polystyrene but not ABS, which is dealt with in the next section; the differences between production and consumption indicate that there are substantial exports but, as there are also imports, do not represent the total amounts exported. The flat period over the years 1954–6 coincides with the development of high impact polystyrene and the start of its availability in commercial quantities. It is interesting to speculate that consumption of the straight polymers had apparently reached its peak at that time but the arrival of the high impact material gave the trend of consumption a substantial upward boost. For the future it appears from the figures that the growth rate is again slowing down. The figures for the growth in production of ABS suggest that this new material has, in effect, taken over the growth of polystyrene and that the output of straight polymer and the conventional high impact materials will only grow slowly in the future.

TABLE 19. U.K. PRODUCTION AND CONSUMPTION
OF POLYSTYRENE
(figures in 000 t/a)

Year	Production	Consumption
1954	23	13·5
1955	22	12·5
1956	25	13
1957	33	18
1958	34	23·5
1959	39	30
1960	42	38
1961	50	42
1962	66	49
1963	74	57
1964	81	62
1965	91	69
1966	104	65
1967	111	70

ABS COPOLYMERS

These materials, in which acrylonitrile, butadiene and styrene
are copolymerized in various ways, were first introduced by the
Naugatuck Chemical Company in the United States in 1948.
Small quantities were imported into this country to develop the
market, largely by Marbon (part of the Borg-Warner Corporation) through their U.K. associates, Anchor Chemical Co. Ltd.
In the early 1960's several companies established manufacturing
facilities in this country, notably Marbon themselves at Grangemouth, Monsanto at Newport, Sterling at Stalybridge, B.P.
Chemicals (U.K.) Ltd. at Barry, Uniroyal at Bromsgrove and
BXL at Manningtree. Capacities are rather indeterminate since
some of the plant is multi-purpose in character; thus, Monsanto
use the same plant as was used for their high styrene content
styrene–butadiene latices. The total manufacturing capacity in
the country is, however, believed to be about 25,000 t/a of which
Marbon, who have a plant built specially for the production of

these resins quite separately from any other activities, are understood to account for 10,000 t/a.

The growth in production of these copolymers in recent years is shown in Table 20.

TABLE 20. U.K. PRODUCTION OF ABS COPOLYMERS

1960	400 tons	1964	8,400 tons
1961	800 tons	1965	10,500 tons
1962	1500 tons	1966	13,000 tons
1963	3,150 tons	1967	14,000 tons

RAW MATERIAL

Manufacture of styrene is described at the beginning of this chapter and the sources of butadiene are dealt with in Chapter 13 as its main outlet is for synthetic rubber.

Acrylonitrile is produced in the United Kingdom by Border Chemicals Ltd. at Grangemouth. This is a joint company of B.P. Chemicals (U.K.) Ltd. and I.C.I. Ltd.; the capacity of the plant is about 40,000 t/a and major extensions have been announced. Plans for a plant of about the same size have also been announced by Monsanto. A process in which propylene is reacted with air in the presence of ammonia over a special catalyst is employed. The reaction may be represented:

$$CH_2 = CHCH_3 + NH_3 + 1\tfrac{1}{2}O_2 \longrightarrow CH_2 = CHCN + 3H_2O$$

Propylene   Acrylonitrile

In practice the reaction is rather more complex and a number of by-products are produced, including significant quantities of hydrogen cyanide.

COPOLYMERIZATION

Several steps are involved in the production of the commercial copolymers; a considerable number of process variations and combinations are in use and are largely kept secret. It appears that there are three main types of process, each with several steps.

1. (a) Butadiene and acrylonitrile are emulsion copolymerized at 40°C for 17 hr, using a peroxide catalyst, and the resulting latex is stripped of any excess butadiene and acrylonitrile.

    (b) Styrene and acrylonitrile are copolymerized at 70°C for 4–6 hr, with persulphate as a catalyst to form a similar latex.

    (c) The two latices are mixed, coagulated, filtered and the precipitated polymer dried.

2. (a) Butadiene is emulsion polymerized (see polybutadiene rubber, p. 274) to form a latex.

    (b) Styrene and acrylonitrile are added to the latex and polymerization continued, a graft copolymer latex being formed in which the styrene–acrylonitrile copolymer chains grow from some of the residual points of unsaturation in the polybutadiene chains.

    (c) Styrene and acrylonitrile are copolymerized to form a similar latex.

    (d) The two latices are mixed and the polymer recovered as in step 1(c) above.

3. Styrene and acrylonitrile are copolymerized in the presence of poly–butadiene emulsion and the polymer recovered as before.

Of the various alternatives, number 3 gives a product with the highest impact strength while 2 gives the highest tensile strength; 1 appears to yield a product of average properties, including good temperature resistance.

The process combination described under 2 is probably the most widely operated in the United Kingdom. The polybutadiene is about 20% of the total, the remainder being about three-quarters styrene and one-quarter acrylonitrile. The product can, perhaps, best be described as a dispersion of polybutadiene grafted with styrene and acrylonitrile in a matrix of styrene–acrylonitrile copolymer. A typical production sequence is as follows.

Butadiene is emulsion polymerized under slight pressure in the presence of a little carbon tetrachloride to act as a chain stopper. When about 75% of the diene has polymerized, a small amount of sodium dithiocarbamate is added to stop the reaction altogether and the latex is steam stripped to recover unreacted butadiene. This latex is then transferred to a second reactor in which styrene and acrylonitrile are copolymerized in the presence of a persulphate or a peroxide. Much of the styrene and acrylonitrile polymerize as side chains on the polybutadiene. For material of the highest impact strength the process is stopped at this stage and the latex is transferred to a flocculation tank where the emulsion is broken; the precipitated ABS material is filtered off, washed and dried to a crumb form. For a product of medium impact but high tensile strength the latex is blended with more styrene–acrylonitrile copolymer and the product finished as above.

PROPERTIES

The properties can be varied widely according to the composition. The tensile strength normally ranges between 2500 and 9000 lb/in^2, the impact strength between 3 and 12 ft-lb/in. of notch, and heat distortion temperatures between 70° and 110°C are obtainable. The highest tensile and impact strengths are not found in the same sample. The material is generally more rigid than the polyolefins but more flexible than polystyrene. Medium impact and tensile strength materials are usually chosen for injection moulding, high tensile material for extrusion and high impact for calendering.

APPLICATIONS

ABS can be used for nearly all products made from thermoplastics and might almost be regarded as the universal thermoplastic; its one great disadvantage is its higher price compared with the older products. The fact that it can be injection moulded as easily as polystyrene has made it a sort of super high impact polystyrene and it is equally suitable for sheet extrusion and vacuum forming. This does not mean, of course, that any sample of ABS has all the desirable properties; it must be formulated for the end use in mind. Thus, the higher proportions of acrylonitrile required to give maximum heat and chemical resistance make a less flexible polymer which is more difficult to process. The end-use breakdown for 1967 is estimated to be as shown in Table 21.

TABLE 21. END-USES OF ABS PLASTICS 1967

	%
Telephones	21
Sheet	20
Consumer durables	20
Automobile (other than Sheet)	10
Pipe	10
Mechanical applications	9
Other extrusions	5
Miscellaneous	5

One of the major consumer durable applications has been for bath tubs. Many applications are using relatively small quantities of ABS where its cost–property relationship is being tested against the more conventional thermoplastics; pipe appears to be a particularly promising outlet.

READING LIST

Styrene, its Polymers, Copolymers and Derivatives, by R. H. Boundy and R. F. Boyer, Reinhold, New York, 1952.
Polystyrene, by Teach and Kiseling, Reinhold, New York, 1960.

CHAPTER 10

Polyvinyl Chloride and its Copolymers

As IN the case of some other plastics, the formation of solid polyvinyl chloride (PVC) from its monomer was first noticed in the nineteenth century but commercial exploitation of the material had to await the development of the necessary techniques for processing it in the second quarter of the twentieth. Once started, development was remarkably rapid, as shown by the figures for U.K. production given in Table 22.

TABLE 22. U.K. PRODUCTION OF PVC
(figures in 000 tons)

1954	42	1961	107
1955	48	1962	115
1956	47	1963	147
1957	68	1964	177
1958	73	1965	194
1959	87	1966	199
1960	100	1967	220

Development began in both the United States and Germany before the Second World War, but it was not until after the fall of Malaya in 1942 that there was any large scale production in this country. The shortage of rubber made the production of substitutes essential and there were many applications of rubber which could be met satisfactorily by plasticized PVC, even though its elastic recovery properties were much inferior to those of natural rubber. A plant to produce both monomer and polymer was set up by the Ministry of Supply at Hillhouse, near Fleetwood, Lancs, which was operated by I.C.I. on behalf of the Ministry.

After the war I.C.I. took over the Hillhouse factory and soon after this the D.C.L. came to an arrangement with the B. F. Goodrich Company of the United States to set up a joint undertaking, British Geon Ltd. (since 1966, B.P. Chemicals (U.K.) Ltd.), to use Goodrich and D.C.L. know-how for the manufacture of PVC at Barry in South Wales. Later, Bakelite Ltd. established a small plant at Aycliffe, Co. Durham. The first two establishments have been steadily expanded until, at the time of writing, their capacities are estimated to be 180,000 t/a and 140,000 t/a respectively.

THE MONOMER

Vinyl chloride, $CH_2{=}CHCl$, is a gas at ordinary temperatures and has a boiling point of $-13 \cdot 8°C$; it must, therefore, be stored and transported in closed and pressurized containers. There are two main processes for its manufacture starting from acetylene or ethylene. From acetylene the reaction is a simple addition of hydrogen chloride according to the equation:

$$CH{\equiv}CH + HCl \longrightarrow CH_2{=}CHCl$$

From ethylene the process is a two-stage one and involves the preliminary conversion of ethylene to ethylene dichloride which is then decomposed by pyrolysis to vinyl chloride and hydrogen chloride.

$$CH_2{=}CH_2 + Cl_2 \longrightarrow CH_2ClCH_2Cl$$

$$CH_2ClCH_2Cl \longrightarrow CH_2{=}CHCl + HCl$$

At one time the acetylene route was the most widely used but, as ethylene from modern high efficiency naphtha crackers has become steadily cheaper, the ethylene route has become more popular. Disposal of the by-product hydrogen chloride from ethylene dichloride pyrolysis was at first a problem and tended to favour the use of both processes side by side. More recently, however, combined oxidation–chlorination processes have been devised which, in effect, oxidize the hydrogen chloride to chlorine and water as it is produced according to the overall equation:

$$2CH_2{=}CH_2 + Cl_2 + \tfrac{1}{2}O_2 \longrightarrow 2CH_2{=}CHCl + H_2O$$

It now seems likely that ethylene will be the preferred raw material of the future. At present there are two producers of the monomer in the United Kingdom: B.P. Chemicals (U.K.) Ltd. who use the two processes side by side at Barry, and I.C.I. who, while still using the two processes separately at Hillhouse and Runcorn, have a major complex under construction at Runcorn which will use an oxychlorination process of their own development.

POLYMERIZATION PROCESSES

As for the polymerization of styrene (see p. 181), emulsion, suspension, bulk and solvent processes may be used. For PVC the first two processes are preferred and are carried out by methods broadly similar to those already described for polystyrene; the differences are great enough, however, to justify a fairly full description.

Emulsion Process

The monomer is emulsified in water with an emulsifier such as sodium lauryl sulphate or ammonium stearate: a water-soluble initiator, e.g. ammonium persulphate, is added and the reaction mixture is stirred vigorously. According to one theory of polymerization vinyl chloride monomer, solubilized in soap micelles, is more readily accessible to free radicals formed from the initiator in the aqueous phase than monomer in emulsified droplets. Accordingly the initial locus of polymerization is the micelle. A polymer particle starts to form, with emulsifier absorbed on its surface, and grows through diffusion of vinyl chloride into it through both the water phase and the emulsifier layer. As polymerization proceeds, more and more monomer is adsorbed by polymer particles until, by the time it is 60% complete, all the remaining monomer is in the polymer phase. These monomer molecules tend to add on to existing chains rather than form new ones since they are practically insulated from the action of freshly formed free radicals. Molecular weight is determined by reaction temperature but, if necessary, it may be controlled by the addition of a chain stopper such as ethylene dichloride. When polymerization is complete the "latex" is broken by the addition of an electrolyte, the precipitated polymer washed in a centrifuge and dried. The polymer particles formed by this process are exceedingly small and slightly porous.

The process may be set up to operate on a continuous basis but batch operation is still widely used.

Suspension Process

In this process liquid vinyl chloride is broken up by agitation into droplets suspended in water containing a protective colloid, such as starch, methyl cellulose or polyvinyl alcohol, which increases the viscosity of the aqueous phase and so decreases the tendency of the droplets to coalesce. Polymerization is initiated in the vinyl chloride phase and an initiator, usually an organic peroxide, soluble in the monomer must be used. As the polymer chains grow the dispersed phase becomes more viscous and particles of polymer separate out to form a separate polymer phase; this causes the rate of polymerization to increase rather than decrease. Polymerization is believed to continue in both the PVC and the monomer phases; the former results from the diffusion of vinyl chloride to active "sites" in the semi-solid polymer where, due to the high viscosity and consequent lack of mobility, chain termination is slowed up.

When agitation is stopped the polymer particles separate out; they are washed to free them from catalyst and protective colloid, partly dried by centrifuging and finished off in a low temperature oven or on a drum drier. The process is invariably operated batchwise; the polymer particles are far larger than those obtained from the emulsion technique. By using special dispersing agents, such as a maleic anhydride–vinyl acetate copolymer, irregularly shaped particles with a very large surface area are obtained. These absorb plasticizer (see p. 207) more easily than the small dense particles obtained with the more conventional dispersing agents so that dry powder mixes can be obtained by simple mixing at temperatures of the order of 100°C.

Bulk and Solvent Polymerization

Bulk polymerization is being used by one French company and a solvent process by one company in the United States. Bulk polymerization is claimed to give a product which is free from surface active agents. According to patents azodi-isobutyronitrile is used as the initiator.

Butane is reported to be the solvent used in solution polymerization and benzoyl peroxide the initiator. The advantage claimed for this process also is that the product is free from emulsifying agents and other additives used in the emulsion and suspension methods. The molecular weight is, if anything, lower than that produced by the bulk method; it is sometimes difficult to get rid of the last traces of the solvent.

Effect of Molecular Weight Differences

In general, the lower the molecular weight the poorer the mechanical properties of the polymer and the worse its thermal stability. Against this, however, the lower molecular weight polymers have better flow characteristics. A moderate molecular weight with a fairly narrow distribution is best for rigid PVC products whereas, for the plasticized material, a wider molecular weight distribution is preferable and there should be at least some low molecular weight material present. Uniformity in particle size is also of great importance as a wide distribution of particle sizes gives poor flow properties. Although the low molecular weight material can be easily worked it tends to be brittle; as the molecular weight increases the material becomes tougher but, eventually, it becomes impossible to work it on conventional plant.

Structure of the Polymer

The vinyl chloride units in the polymer molecule are mainly linked in a uniform head to tail sequence:

$$-CH_2-CH-CH_2-CH-CH_2-CH-$$
$$\ \ \ \ \ \ \ \ |\ \ \ \ \ \ \ \ \ \ \ \ \ |\ \ \ \ \ \ \ \ \ \ \ \ \ |$$
$$\ \ \ \ \ \ \ \ Cl\ \ \ \ \ \ \ \ \ \ \ Cl\ \ \ \ \ \ \ \ \ \ Cl$$

Spectrographic studies have shown that there are about 16 side chains per molecule of polymer but these are mainly short so that the molecule is effectively a straight chain. Commercial PVC is mainly amorphous but small areas of crystallinity do occur where the molecule appears to be syndiotactic.

Use of Ziegler–Natta catalysts, of the type described in Chapter 8 on polyolefins, to give a more regular structure is not satisfactory; yields are small and the catalyst residues, which are difficult to remove, accelerate heat degradation. There is some evidence that formation of syndiotactic polymers is favoured by polymerization at low temperatures, but little commercial use has been made of this as the polymer has a high softening point without any increase in the thermal stability.

There is evidence that heat stability increases with molecular weight but the high molecular weight polymers would be very difficult to process.

Processing of PVC

It is probably more important for PVC than for any other plastic that the additives and the processing conditions used should be chosen specially for the end use for which the material is designed. The polymer, as produced, is heat sensitive and difficult to work and, although unplasticized products consisting virtually of 100% PVC are now made on a large scale, the PVC industry of today could not have been developed without the plasticizers which have been simultaneously produced to meet its requirements. If it is realized that total U.K. usage of PVC is approaching 200,000 tons annually, of which the major part will be mixed with up to 50% of plasticizer, it will be appreciated that the manufacture of these compounds is almost a special branch of the chemical industry.

It is also worthy of note that, probably because many of the early applications of PVC were as a substitute for rubber, the processing techniques applied to it are somewhat similar to those of the rubber industry.

UNPLASTICIZED COMPOUNDS AND PRODUCTS

An unplasticized compound normally contains about 95% of polymer plus a small quantity of lubricant, such as a stearate, suitable dyes or pigments and a heat stabilizer. Mixing of the additives can be carried out in Werner–Pfleiderer or Bridge–Banbury type mixers and the compound is usually extruded in the form of sheets or tubes. The extrusion conditions are most critical as the softening point of the compound is usually only 20–30°C below the decomposition temperature and, for this reason, small quantities of a plasticizer or softener may be added. These have little effect on the rigidity and do make processing easier but they can greatly reduce resistance to chemical attack, for which unplasticized PVC is unsurpassed by any other plastic material within its price range. For the extrusion of unplasticized material, emulsion process polymer is often preferred as the particles are slightly easier to process and the traces of emulsifier on the surface assist extrusion.

Rigid PVC pipes are used for transporting oil and water over long distances and are sometimes used in chemical plants where resistance to severe corrosive conditions is required; they are also finding a large and growing outlet in the building industry for drainage and soil systems, guttering and rainwater pipes and similar applications. They are not, however, recommended for use above 70°C for long periods and cannot, therefore, be used for hot water service. Chlorinated PVC (see p. 214) has higher temperature resistance combined with excellent resistance to chemicals but is, of course, more expensive.

PVC compounded with small amounts of special rubbers and processing aids, in addition to being more easily worked, has a higher impact strength than the unplasticized material although it is still rigid. Sheets are readily produced by calendering and can be vacuum formed into a variety of articles such as advertising signs, refrigerator parts, luggage and containers of many kinds.

It must be emphasized that a great volume of work has been done on the production of rigid grades of PVC, most of which is unpublished. Particle size, molecular weight distribution, small quantities of additives and detailed processing methods all play their part in producing the right characteristics for a particular application and each manufacturer tends to develop his own "know-how" as part of his stock-in-trade.

Plasticized PVC

As already noted, by far the major proportion of PVC is plasticized and, for most applications, the plasticizer added is at least 30% by weight of the polymer. Since the plasticizer is mechanically and not chemically bound in the compounded material, it will be subject to loss by evaporation and migration and must have a low vapour pressure and be chemically inert. The large quantity required makes cost of major importance.

Most plasticizers for PVC are high boiling esters; one of the first used was tritolyl phosphate and this was followed by the esters of phthalic acid which still make up the great bulk of the plasticizers of commerce. Although phthalic acid has remained as the acid from which most plasticizers are made, other dibasic

acids, such as sebacic and adipic, have become freely available
and are used to some extent although they increase the cost of the
ester. There have been changes also in the alcohol component,
the tendency being towards the use of progressively higher boiling
alcohols to obtain esters of the lowest possible vapour pressure.
n-Butanol was one of the first alcohols used and much dibutyl
phthalate went into the early PVC compounds. It has the
advantage of being reasonably cheap, of compounding easily and
of giving products with excellent flexibility over a wide range
of conditions. It is, however, comparatively volatile and has a
distinct odour, so that products containing it not only have a
marked odour but also gradually lose their plasticizer and become
brittle. Its use was one of the reasons why PVC tended to be
regarded as only a substitute for "the real thing" in the early days
of its development.

Reasonably cheap higher alcohols have now become available,
especially from the petroleum chemical industry, and modern
plasticizers are based largely on octyl and nonyl alcohols from
various raw materials. These high boiling alcohols are also becoming important for synthetic detergents and their production
has become an important branch of the chemical industry; for
further details the student is referred to the reading list at the end
of this chapter.

In addition to the esters mentioned above, some polymeric
materials are used as plasticizers and there is a large group of
compounds, generally called secondary plasticizers, softeners or
extenders, which are used to cheapen the mix, usually at the
expense of some deterioration in the properties of the final
product.

EFFECT OF DIFFERENT PLASTICIZERS

The plasticizers briefly described above may be divided into
four groups:

(i) Organic esters—e.g. the phthalates and the more expensive adipates and sebacates.

(ii)   Esters of inorganic acids—e.g. tritolyl or trixylenyl phosphate.

(iii) Polymeric materials such as poly (propylene glycol sebacate).

(iv)   Secondary plasticizers and extenders, of which high boiling aromatic hydrocarbons and chlorinated paraffins are typical examples.

The choice of type and quantity of plasticizer to be used is a compromise which seeks to obtain minimum cost with maximum desirable properties. The possibilities are obviously numerous and only a few generalizations can be given here. The esters from group (i) are the most widely used, often in combination with a selected material from group (iv). The inorganic esters are used to impart strength and fire resistance at the expense of flexibility at low temperatures. The polymeric group, which are generally the most expensive, are used mainly where the product is likely to come into contact with fats or vegetable oils, particularly with foodstuffs, and it is important that the plasticizer should not migrate into the fat or oil. In general, increasing the quantity of plasticizer gives a softer and more flexible product.

Precise comparison of plasticizers is only meaningful when the formulations and other conditions are carefully controlled and is mainly useful when comparing formulations for a specific application. In general it can be said that, at room temperature and at similar proportions, dibutyl phthalate, followed by dialphanol and di-2-ethylhexanol phthalates gives the most flexible compounds while dinonyl phthalate gives the stiffest. (Alphanol is a mixture of C_7–C_9 primary alcohols made from C_6–C_8 α olefins.) Tritolyl phosphate gives the highest tensile strength product and the sebacates and adipates the lowest; the phthalates are intermediate. The sebacates give the maximum elongation at break and dibutyl phthalate the minimum.

When the plasticizers are compared at quantities which give equal degrees of softness, dinonyl phthalate gives products with the highest electrical resistance, closely followed by the other phthalates, while the phosphates and sebacates give the lowest.

Polymeric plasticizers from group (iii) are only used on a small scale in applications where low toxicity and freedom from migration are the required characteristics; in addition to the polyester type already mentioned a liquid butadiene–acrylonitrile copolymer is sometimes used.

MECHANISM OF PLASTICIZATION

The function of a plasticizer is believed to be to penetrate between the polymer chains and hold them apart. When cold, PVC chain molecules are held tightly together by intermolecular forces and the material is hard and inflexible but, on heating, the chains acquire sufficient energy to become disentangled. Under these conditions the plasticizer molecules can penetrate the space between the polymer chains and prevent them coming together again on cooling so that the cooled product remains soft and flexible. Since the plasticizer is only a physical mixture, with the polymer, it can evaporate or be extracted with solvents, when the PVC will revert to its original hard and inflexible state; this is why there has been a constant search for plasticizers of very low vapour pressure which are both compatible with PVC and resistant to extraction by the materials with which the finished articles will come into contact.

It is possible to produce softer polymers by copolymerizing vinyl chloride with a monomer such as 2-ethylhexanol acrylate or diethyl maleate. In this case the material is sometimes said to be internally plasticized, and volatilization and migration of plasticizer is no longer possible. There is, however, some diminution of other desirable characteristics and low temperature flexibility is only fair.

Stabilizers

All PVC compounds require stabilization against heat, oxidation and, where appropriate, against the effects of light. Heat stabilizers are of the utmost importance since degradation starts at the temperatures at which the material is normally processed and, once started, it is difficult to stop. Thermal degradation is accompanied by emission of hydrogen chloride and has the effect of discolouring the product and of making it weaker. Light, particularly ultraviolet light, has a deleterious effect on PVC and there appears to be an induction period after which degradation takes place by atmospheric oxidation; these two effects are, to some extent, interdependent.

There are broadly three kinds of stabilizer—hydrochloric acid acceptors, antioxidants and ultraviolet light absorbers. Hydrochloric acid acceptors, which protect the material against heat degradation, are usually metallic soaps such as the stearates of calcium, barium, cadmium and lead; most insoluble lead salts are effective. Stearates have lubricating properties and also act as processing aids but they may give trouble with "blooming" in quantities above 2% on the PVC. Organic tin compounds are useful; the mercaptides are especially good and compounds such as dibutyltin dilaurate are also used. Epoxy resins, though not particularly effective when used alone, have a synergistic effect in conjunction with a stearate or laurate and are, therefore, especially useful where a non-toxic stabilizer is essential. Usually 1–2% of thermal stabilizer, calculated on the PVC content of the compound, is adequate but, in exceptional cases, higher quantities may be used.

Many of the heat stabilizers in common use have some antioxidant properties and it is not usually necessary to add phenol or amine type antioxidants as well. Organic phosphates used as plasticizers have antioxidant properties and show heat stabilizing properties as well, especially when used in conjunction with stearates.

Light stabilizers are added whenever the product has to remain translucent or to retain its colour in strong sunlight; the most useful compounds are hydroxybenzophenones or benzotriazole.

Fillers

The use of fillers in PVC compounds is one method of reducing the cost of the product and is sometimes an alternative to the use of extenders referred to on p. 208; whiting or clay are most commonly used. They give a slightly stiffer material and extra plasticizer may be required to maintain the same degree of flexibility with a consequent loss of tensile strength. Where flexibility is not of great importance, in floor coverings for example, substantial loadings of filler, even up to 30% of the polymer, can be used. In compounds for cable insulation the inclusion of 5–10 parts of clay per 100 of polymer actually improves the insulating properties.

Pigments

PVC compounds may be produced in a wide range of colours, including metallic effects, by the addition of suitable dyes or pigments but the colouring matters must be carefully chosen and must themselves be stable at processing temperatures and to light and must not react with other components of the compound. Pigments containing manganese, cobalt, copper and, above all, iron and zinc, must be avoided. The first three are oxidation catalysts and the last two react with the product. It is sometimes possible to combine pigmentation with stabilization; for example, cadmium and lead compounds can have a dual function. Metallic effects can be obtained by the addition of aluminium powder but this severely limits the amount of filler which can be added if a matt finish is to be avoided.

Lubricants

Without lubricants the compounds tend to stick to hot processing equipment. This can be avoided by the addition of fatty acid salts; as with pigments, some suitable compounds can have a dual function; for example, some stearates may be lubricant, heat stabilizer and antioxidant at the same time.

Manufacture of Plasticized PVC Compounds

The solid ingredients are usually mixed dry in Banbury or Werner–Pfleiderer type mixers and the liquid plasticizer is then added slowly, for example by spraying it in, while the mixing continues. The premix, now in the form of a dough, is fluxed on rolls at about 150°C, which produces a homogeneous sheet that can be removed, cooled and disintegrated into a form suitable for sale or for transfer to an extruder calender.

The stability of a finished PVC article depends largely on its heat history and mixing procedures have been developed to produce compounds which have not been fluxed on rolls and which have not been subjected to temperatures above 110°C at any stage in processing. As noted on p. 204, a specially porous polymer can be produced which readily absorbs plasticizer at temperatures of 95–105°C and gives a free-flowing powder of the same particle size as the original polymer. The restriction of processing temperature gives products with better colour retention and electrical properties; compounds made in this way are not, however, suitable for calendering.

Paste Forming

A compound in the form of a viscous liquid is advantageous for a number of applications such as slush moulding (rotary casting) and spraying techniques. A paste-forming grade of polymer can be produced by spray drying a PVC emulsion. Under appropriate conditions non-porous particles of the rather critical size distribution required are obtained. When this is mixed with plasticizers, stabilizers, pigments, etc., in the cold, the polymer forms a fluid dispersion in the liquid plasticizer; this is known as a plastisol. On heating to about 150°C the plasticizer is readily absorbed by the PVC to form a solid gel; this operation is known as gelling. For best results it is important that the particles should vary considerably in size; if they are too uniform in size they cannot pack together and high paste viscosities are encountered.

The required size variation is sometimes achieved by the inclusion of a small proportion of suspension polymer in the mix.

Manufacturing Problems

PVC differs from other plastics in that much of the formulation and processing of compounds is carried out by the manufacturers of finished articles. In practice the success of a particular application may well depend on the development of an individual formulation and processing technique for it. In consequence, each manufacturer develops his own methods and formulae and, while there is a pool of general information, much individual skill and knowledge is spread over a large number of manufacturers.

MODIFIED PVC AND COPOLYMERS

Many attempts have been made to modify PVC to make it easier to process, to give it improved heat stability or to widen its field of application. As well as modification of the polymer itself, these include copolymerization with other monomers, and copolymers with some unsaturated esters have already been mentioned (p. 226). Others are described below.

Chlorinated PVC

It is possible to chlorinate PVC by passing chlorine into an aqueous dispersion of the polymer; the latex from an emulsion polymerization is suitable for this purpose and the reaction is carried out at about 60°C under the influence of ultraviolet light. The chlorine merely replaces some of the hydrogen in the chain.

$$—CH—CH—CH_2—CH—CH—CH—CH_2—CH—CH_2—$$
$$\quad |\quad\ |\qquad\quad\ |\quad\ |\quad\ |\qquad\quad |$$
$$\quad Cl\quad Cl\qquad\quad Cl\quad Cl\quad Cl\qquad\quad Cl$$

Normally enough chlorine is allowed to react to raise the density from 1·4, the figure for normal PVC, to about 1·6. The products have improved thermal stability and high impact strength and can be compounded and processed in the usual way. Unfortunately processing is more difficult than when using ordinary unplasticized PVC but it can be extruded and is therefore, used mainly for pipe; this pipe is eminently suitable for chemical plant applications where good resistance to chemical attack and reasonably high temperature resistance are required. It can also be used for gas piping.

Expanded PVC

Small quantities of PVC foams are made at present and it has been estimated that these will reach the 1000 t/a mark by 1970. Open cell flexible foams are made by dissolving an inert gas in a plastisol under pressure (p. 213) and allowing the mixture to escape through a nozzle at atmospheric pressure. Foaming takes place and the mixture is gelled by heat. Alternatively a blowing agent which evolves an inert gas on heating can be used with a plastisol. Many blowing agents have been patented including hydrazine mononitrate, carbonyl azides and azodiamide oximes. Closed cell flexible foams are made with a plastisol and a blowing agent foamed in closed moulds. These flexible foams have not the elastic properties of polyurethane or synthetic rubber foams but have advantages in chemical and fire resistance.

Rigid PVC foams are made by similar techniques but foaming takes place under high pressure conditions, usually in the presence of additives such as halogenated hydrocarbons or, in some cases, other polymers.

PVC COPOLYMERS

Vinyl acetate and vinylidene chloride are the monomers most commonly copolymerized with vinyl chloride. Vinyl acetate is made by B.P. Chemicals (U.K.) Ltd. from acetic acid and acetylene, by British Celanese Ltd. from acetaldehyde and acetic anhydride and by I.C.I. Ltd. using a more modern process in which ethylene is oxidized under controlled conditions in the presence of acetic acid. Vinylidene chloride is made in the

United Kingdom by I.C.I.; the usual method of manufacture is by dehydrochlorination of trichloroethane with the aid of a lime slurry; it may also be made by pyrolysis of trichloroethylene.

Most companies making PVC also produce vinyl acetate copolymer, the vinyl acetate component varying between 5 and 20%. The large acetate groups keep the chains further apart and cause the polymer to have a lower density, lower softening temperature and reduced hardness and stiffness when compared with a straight vinyl chloride polymer. One great advantage is that even a small proportion of vinyl acetate allows the polymer to be processed at temperatures further removed from its decomposition temperature. The products are water resistant and dimensionally stable but are attacked by, or partly soluble in, many organic solvents.

Suspension or emulsion polymerization techniques can be used and some commercially important vinyl chloride–acetate copolymers are solution-polymerized in solvents such as butane or toluene in which the monomers but not the polymer are soluble. Compounding is similar to that for PVC but less plasticizer is required to give the same degree of softness; much of the product is, however, used unplasticized.

Low molecular weight copolymers are used in ketone solution for stoving coatings in the metal decorating field. For injection moulding of unplasticized material a molecular weight of about 10,000 and a vinyl acetate content of 10–15% give excellent results; for compression moulding a slightly higher molecular weight is preferred. The moulded products show low shrinkage and are largely used for gramophone records. Sheet can be extruded from higher molecular weight material (about 15,000) containing about 10% of vinyl acetate. This is easier to extrude than 100% PVC sheet but still has excellent water resistance.

Copolymers of vinyl and vinylidene chlorides in which there is only a small quantity of the latter do not differ very much from the corresponding vinyl acetate copolymer. They are now made only on a small scale in the United Kingdom but have proved useful in calendering and extrusion where low processing temperatures are required. There is, however, an important copolymer containing about 85% of vinylidene chloride, the well known "Saran" series produced by Dow in the United States. The vinylidene chloride polymerizes much faster than vinyl chloride and, by varying the polymerization conditions, substantial variations in molecular weight, molecular weight distribution and arrangement of the vinyl chloride–vinylidene chloride units in the polymer chain may be brought about. This produces a range of materials with softening points which may vary from 70° to 180°C and which, unplasticized, may have varying degrees of stiffness from flexible to rigid. The polymers containing a high proportion of vinylidene chloride are resistant to all organic solvents and, at room temperatures, they are inert to all acids and alkalis other than strong ammonia but are attacked by halogens and strongly basic organic amines. Stabilizers are necessary to give good resistance to heat and sunlight. Saran is mainly extruded as pipe for use where corrosion resistance is important or formed into film by blow moulding; the film has a very low moisture transmission rate. It may also be extruded through an orifice to form fibres. "Saran" is not made in the United Kingdom but is imported on a modest scale, mainly from the United States; its high cost has prevented its development on a large scale.

END USES FOR PVC

PVC is so widely used that a detailed description of its applications cannot be attempted in the limited space available. The figures for the estimated consumption of PVC in various outlets, published each year in the January number of the journal *British Plastics*, give an excellent summary and are reproduced in Table 23 for the years 1965, 1966 and 1967.

In addition there is believed to be 3000–4000 t/a consumed in the manufacture of miscellaneous products not listed above and about 7000–8000 t/a used in the production of compounds which are subsequently exported. The figures include the PVC in

TABLE 23. END USES FOR PVC IN THE UNITED KINGDOM
(figures in 000 t/a)

	1965	1966	1967
Rigid			
Extruded (mainly pipe)	27	30	35
Blown bottles	—	1	1
Gramophone records	8	9	8
Sheet	8	7	7
Total rigid	43	47	51
Flexible			
Calendered sheet (e.g. for curtains, table-cloths, protective clothing, decorative material, etc.)	35	35	38
Fabric coating (leathercloth, decorative material)	12	13	17
Flooring	27	28	31
Conveyor belting (mainly for coal mines and some food conveyors)	6	5	5
Miscellaneous extrusions (including flexible tubing)	12	13	13
Electric cables	35	37	37
Dipping and slush moulding (toys, etc.)	8	9	10
Footwear	4	4	7
Total flexible	139	144	158
Overall total	182	191	209

copolymers where PVC is the major component but the border-line here is obviously somewhat uncertain. The totals, however, agree fairly well with the figures for total production given in Table 22 on p. 200.

READING LIST

Vinyl Polymerization, by G. E. Ham, Arnold, London, 1967.
PVC Technology, by W. S. Penn, Maclaren, London, 1962.
The Stabilization of Polyvinyl Chloride, by F. Chevassus and R. de Broutelles, Arnold, London, 1963.

CHAPTER 11

Miscellaneous Thermoplastics

THERE are several thermoplastic materials in general use that, for one reason or another, do not justify a chapter to themselves and which do not fit in readily with the materials already described. Some of them are produced in large quantities but find their main uses in other fields, such as textiles, while others are used mainly as plastics but production is relatively small by plastics industry standards. These products are described in the various sections of this chapter, which may be regarded as the "bit box" of the book; most of them are thermoplastics but the silicone polymers, which may be thermoplastic or thermosetting and, in some forms, are used as elastomers, have also been included.

POLYMERS FOR MAN-MADE FIBRES

1. *Nylon.* Polyamides of the nylon type are now manufactured in very large quantities for synthetic fibre production; about 3000 t/a, however, are currently being used in the United Kingdom for injection moulding and the quantity is increasing. As a plastic, nylons are very tough, have high resistance to abrasion and chemical attack and an exceptionally low co-efficient of dry friction; they are used specially for manufacture of small bearings and gears.

There are two kinds of nylon in general use; nylon 6·6 is a straight chain polyamide formed by reaction between hexamethylene diamine and adipic acid.

$$H_2N(CH_2)_6NH_2 \pm HOOC(CH_2)_4COOH \longrightarrow$$
$$H_2N(CH_2)_6NHCO(CH_2)_4COOH + H_2O$$

nylon salt

$$n\ H_2N(CH_2)_6NHCO(CH_2)_4COOH \longrightarrow$$
$$-[-HN(CH_2)_6NHCO(CH_2)_4CO-]-_n + n\ H_2O$$

The numbers refer to the number of carbon atoms in the two reactants. Other diamines and di-acids may be used.

Nylon 6 is also a linear polymer formed by ring opening and polymerization of caprolactam.

NH'———CH₂———CH₂
  | \CH₂
CO ———CH₂———CH₂   ⟶ $[-NHCO(CH_2)_5-]_n$

Caprolactam   Nylon 6

Both nylons are highly crystalline with high and sharp melting points so that careful temperature control of injection moulding equipment is essential: a moulding temperature approaching 280°C is required for nylon 6·6 and a rather lower temperature for nylon 6.

SATURATED POLYESTERS

The most important saturated polyester is a linear polymer formed by reaction between ethylene glycol and dimethyl terephthalate; like the nylons it is used mainly for textiles and is widely known under the I.C.I. trade name of "Terylene".

The polymer is highly crystalline, has a sharp melting point and a low viscosity in the liquid state. It is unsuitable for injection moulding but may be extruded as a film which, after stretching and heat treatment, has exceptional strength and clarity. It is produced in significant quantities by I.C.I. and sold under the trade name of "Melinex".

POLYVINYL ALCOHOL
AND ASSOCIATED MATERIALS

There is an important group of polymers, containing the repeating unit —CH_2CHR—, which are usually classified under this heading and which may be regarded as derivatives of vinyl alcohol. It may be noted that the three major thermoplastics, polyethylene, polystyrene and PVC, described in Chapters 8–10, are also vinyl polymers.

Polyvinyl Alcohol

Vinyl alcohol has never been isolated and attempts to produce it result in its tautomer, ethylene oxide.

$$CH_2{=}CHOH \longrightarrow CH_2{-}CH_2$$
$$\diagdown\diagup$$
$$O$$

Polymers of the alcohol may, however, be produced by hydrolysis of polyvinyl acetate, described in the next section. Polyvinyl acetate is allowed to swell in anhydrous methanol, $0 \cdot 5 \%$ of caustic soda or hydrochloric acid is added and the mixture gently heated to saponify the acetate to the alcohol. It is difficult to complete the hydrolysis and most samples of the alcohol contain small amounts of the acetate; as the acetate content decreases water solubility, resistance to oils and softening temperature all increase. Polyvinyl alcohol can only be partially re-esterified, indicating that some change in structure—possibly some internal etherification—has taken place.

Unplasticized polyvinyl alcohol does not melt but softens to a rubber-like material when heated; it can be plasticized with glycerol or other high-boiling, water-soluble organic compounds and it can then be cast into films which are very impervious to gases and to oils and greases.

Most polyvinyl alcohol is used in aqueous solution when it has many of the properties of a solution of starch. It is used to increase the wet strength of paper in, for example, paper towels, but the additional strength soon falls; it is also an excellent adhesive and emulsifying agent and has many applications in the formulation of shaving cream and other toiletries.

POLYVINYL ACETATE

Commercial production of polyvinyl acetate started in Canada in 1920, and in 1930 in the United States. Manufacture of the monomer is described in Chapter 10, p. 215.

Vinyl acetate is used as a comonomer with vinyl chloride (see p. 216) but is also suspension or emulsion polymerized alone to give what is, in effect, an aqueous dispersion of swollen polymer particles. Redox catalyst systems (p. 25) are often employed thus allowing low polymerization temperatures to be used and avoiding saponification of the vinyl acetate which is favoured by high temperature. Alternatively a solvent polymerization process may be used with methanol or ethyl acetate, in which the polymer is insoluble, as the solvent. Relatively low molecular weights of 5000–20,000 are reached and the polymers are quite low melting and soluble in most solvents.

Current U.K. production of polyvinyl acetate and copolymers with a high proportion of polyvinyl acetate is about 30,000 t/a mostly in the form of an emulsion with 50–55% solids content.

Uses

By far the most important uses of polyvinyl acetate are as emulsion paints and adhesives, some 50% of total production going to the former application and 30% to the latter; a plasticizer such as dibutyl phthalate is usually incorporated in the emulsion for either application. The resins are thermoplastic so that no heat is required; the water is eliminated from the emulsion by absorption and evaporation so that the swollen polymer particles coalesce to form a coherent film, which adheres strongly to most surfaces. The material is particularly good as an adhesive for porous surfaces such as paper, leather and wood but is also used for bonding other plastics and even for concrete. Some adhesives are made by dissolving the solid resin in an organic solvent.

The solid resin is not used for moulding due to its low softening point and to its tendency to stick in the mould. It is, however, mixed with fillers and used in the manufacture of floor tiles, pressed wood compositions and similar products; it is also the starting point for manufacture of polyvinyl alcohol and acetals.

Some interesting new copolymers of vinyl acetate and ethylene are under development in the United States but are not yet made in the United Kingdom; there is not much information on details of production available. It is believed that the copolymers are made by a high pressure process using about 10% of vinyl acetate in the mixed ethylene–vinyl acetate gas stream. The higher the proportion of vinyl acetate in the polymer the more flexible it becomes; dispersions of a copolymer containing 30% of vinyl acetate show great promise for paper coating.

The copolymers can be processed in the same way as polyethylene but at lower temperatures; they will not stand up to temperatures above 60°C for long periods. They are tough, flexible, have excellent electrical properties and are more resistant to light than 100% polyethylene. Resistance to acids, alkalis, detergents and alcohols is good but the materials are attacked by other organic chemicals. Their main uses seem likely to be for tubes, hose, sheathing, gaskets, disposable gloves and especially for packaging films. So far the materials are not widely used in the United Kingdom and the applications are really still in the development stage.

POLYVINYL ACETALS

These products may be made by condensation of polyvinyl alcohol with the appropriate aldehyde, according to the general equation for the production of acetals:

$$2ROH + RCHO \longrightarrow R'CH \begin{matrix} \diagup OR \\ \diagdown OR \end{matrix} + H_2O$$

In practice the polyvinyl acetals are usually made direct from the acetate, without intermediate isolation of the alcohol.

Polyvinyl–formal is exceptionally tough, with a tensile strength of about 20,000 lb/in^2; it is soluble in water but very resistant to hydrocarbons. One of its earliest uses was, in combination with phenol–formaldehyde resins, for sticking clutch linings to their plates for tanks during the Second World War. This combination formed the basis of the "Redux" adhesives marketed by C.I.B.A.

The higher polyacetals are insoluble in water and are chemically inert. They have found a limited application as plasticizers in cellulose nitrate dopes. Polyvinyl butyral is widely used in the production of laminated safety glass for windshields. The resin is deliberately manufactured with 18–20% of free OH groups, which improve its adhesion to glass; it is normally plasticized with, for example, dibutyl phthalate and is produced as a thin sheet which is used to bond sheets of glass together.

ACRYLATE POLYMERS

This is an important group of polymers which may be regarded as derivatives of acrylic acid, $CH_2{=}CHCOOH$. The free acid may be made by simultaneous hydrolysis and dehydration of ethylene cyanhydrin or by the Reppe process of treating acetylene with carbon monoxide (as nickel carbonyl) and water.

$$HOCH_2CH_2CN + H_2SO_4 + H_2O \longrightarrow CH_2{=}CHCOOH$$
$$+ NH_4HSO_4$$

$$CH{\equiv}CH + CO + H_2O \longrightarrow CH_2{=}CHCOOH$$

If either reaction is carried out in the presence of an alcohol instead of water the corresponding ester is produced. The aldehyde, acrolein, may be produced by controlled oxidation of propylene:

$$CH_3CH{=}CH_2 + O_2 \longrightarrow CH_2{=}CHCHO + H_2O$$

As noted on p. 196, when this reaction is carried out in the presence of ammonia, acrylonitrile is produced directly. Border

Chemicals Ltd. have announced a further modification of the process which permits direct production of acrylic acid esters, but construction of a large scale plant in the United Kingdom has been postponed. Virtually the same process is used commercially in the United States.

Acrylonitrile is very important for the production of textile fibres and its copolymers with styrene and butadiene have been described in Chapters 9 and 13; its homopolymers are not, however, used as plastics. The acrylic ester polymers may be divided into two groups—those based on the esters of acrylic acid and those based on esters of the homologue, methacrylic acid $CH_2 = C(CH_3)COOH$. The latter, although generally classified with the acrylics, differs from them so much in method of preparation and properties that they might well be regarded as a separate group.

ACRYLIC ACID ESTERS

This is a rapidly growing group of polymers; they were first produced in the United States in 1931 but made little progress in this country until after the Second World War. United Kingdom production is now substantial and is currently in excess of 10,000 t/a. There are several manufacturers, some of them relatively small. The monomers are now most commonly produced by the Reppe process and it is believed that Lennig Chemicals Ltd., the British subsidiary of the American Rohm & Haas Company and one of the largest U.K. producers, use this route.

Polymerization

The polymers are manufactured by a conventional emulsion polymerization technique using a peroxide catalyst. Limited quantities of solid polymers have been made by a bulk polymerization method but they have found little commercial use. Polymethylacrylate is tough and rubbery but polymers made from esters with ethyl and higher alcohols are softer and tackier.

Uses

The main uses for the acrylates lie outside the plastics industry. Methyl acrylate is used as a comonomer with acrylonitrile in the production of an important class of fibres, its function being to improve the dyeing properties.

The other acrylates are used as comonomers with vinyl acetate in production of emulsion paints. Polyacrylic ester emulsions are also used in the finishing of leather as a base for cellulose nitrate lacquers, for making textiles resistant to dry cleaning, for coating paper to act as an adhesive for clays where specially white paper is required and for producing washable wallpaper. Their main plastics use is as comonomers and internal plasticizers in manufacture of vinyl polymers, especially PVC.

POLYMETHYLMETHACRYLATE

This is the only polymer of commercial importance derived from methacrylic acid and has been made in the United Kingdom by I.C.I. since the early 1930's. Total U.K. output is of order of 25,000–30,000 t/a.

The monomer is manufactured by first reacting acetone with hydrogen cyanide to form acetone cyanhydrin which is then subjected to successive steps of hydrolysis, esterification and dehydration to give the methyl methacrylate monomer.

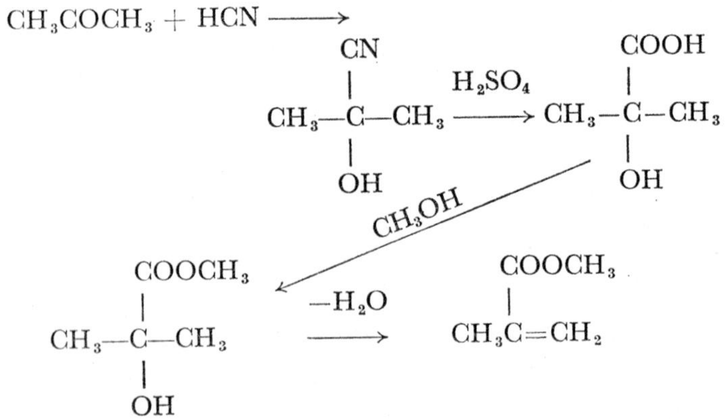

$$CH_3COCH_3 + HCN \longrightarrow$$

Polymerization

A large part of the output of polymethylmethacrylate is produced in the form of relatively thick sheets, for which a bulk polymerization technique is most suitable. A favoured method is to use a casting syrup, made either by partially polymerizing the monomer in the presence of an initiator or by dissolving polymer in monomer to give a solution of suitable viscosity. About 4% of dibutyl phthalate or n-butyl acrylate is added as plasticizer and the syrup is poured into a cell made of two sheets of heat resistant glass separated by a flexible gasket; there is considerable shrinkage during the final polymerization stage and the gasket must compress sufficiently to take this up. The final curing takes place in an autoclave in an inert atmosphere of carbon dioxide or nitrogen to avoid the formation of bubbles. After cooling, the glass plates are separated and re-used and the polymethylmethacrylate sheet is annealed; scrupulous attention must be paid to the cleanliness of the glass plates if sticking is to be avoided.

The finished sheet is transparent with quite outstanding optical properties, while it has a tensile strength of 7000–9000 lb/in^2. Its impact strength is only about $0 \cdot 5$ ft-lb/in. notch, which is low compared with other thermoplastics except straight polystyrene, but is far higher than glass, which the polymer has replaced for many applications, such as aeroplane windows, cockpit screens and radar enclosures. Poor scratch and abrasion resistance have limited its use in motor cars and in building. The sheet is widely known in the United Kingdom under the I.C.I. trade name of "Perspex".

In addition to its use for transparent sheet about one-third of the output of polymer is made into moulding powder. In this case a suspension polymerization method is used; benzoyl peroxide is a typical catalyst, sodium polyacrylate is used as a suspending agent and a little stearic acid as a lubricant. The monomer boils at 80°C and the polymerization is carried out at 100°C for about an hour under slight pressure. The beads are separated by filtration or in a centrifuge and dried in a fluidized bed with heated air.

The beads are then usually mixed on rolls with the appropriate dyestuffs and about 2% of dibutyl phthalate as plasticizer. The resulting product is ground to form the moulding powder, which may be processed by extrusion or injection moulding. Relatively high pressures are required, due to the high viscosity of the molten polymer and one of the functions of the dibutyl phthalate is to improve the flow properties. Polymethylmethacrylate moulding powders are made and sold in the United Kingdom by I.C.I. under the trade name of "Diakon".

Uses

Polymethylmethacrylate sheet is particularly suitable for thermoforming and, in addition to its uses in aircraft mentioned above, it has many applications in the production of lighting equipment, advertising signs and similar applications where its transparency can be exploited; for some of these applications injection moulding may be an alternative to thermoforming. The moulded and extruded products are used mainly for decorative purposes and all kinds of fancy articles are made. It is relatively expensive as moulding powders go and is used only where its appearance and optical properties justify the extra cost.

A limited amount of polymethylmethacrylate, usually in combination with other polymers in solution, is used in surface coating applications, especially to give a very high gloss for motor vehicle finishes.

POLYCARBONATES

The polycarbonate resins, developed commercially by Bayer in Germany and by General Electric in the United States, first appeared on the market in 1959. They may be regarded as linear polyesters of carbonic acid with a dihydroxy compound, of the general formular

$$-\left(-OROC-\right)_n$$
$$\qquad\ \ \|$$
$$\qquad\ \ O$$

For commercial polycarbonates diphenylol propane is the only dihydroxy compound used and is converted to a polycarbonate resin by reaction with carbonyl chloride in the presence of pyridine.

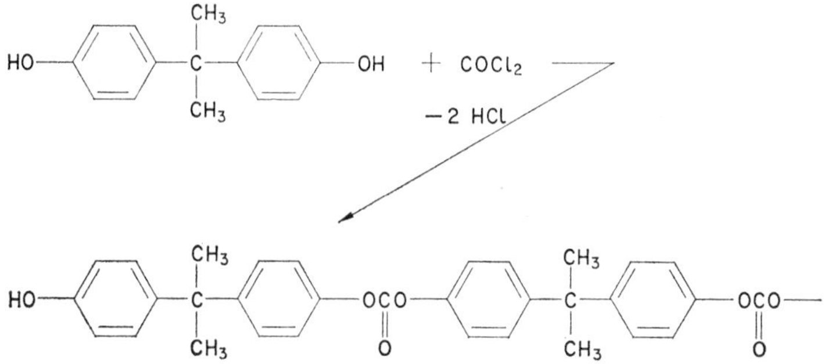

The polycarbonates are hard, tough, rigid and thermoplastic and have a density of $1 \cdot 2$; they can be made virtually transparent with the high impact strength, for a rigid polymer, of 2–3 ft-lb/in. notch. Their tensile strength at room temperature is exceptionally high, 8–10,000 lb/in^2 but this falls progressively with rising temperature. The polymers can stand up continuously to temperatures of 120°C, the heat distortion temperature being about 15°C higher, but temperatures of 250–300°C are required for processing.

Polycarbonates can be compression moulded, injection moulded, extruded, blow moulded or cast while mechanical finishing operations such as turning, milling and drilling are easily carried out. Their chemical resistance is variable; they are very resistant to water and stand up well to dilute acids, oxidizing agents, vegetable oils and aliphatic hydrocarbons but are attacked by alkalis and most oxygenated and chlorinated organic compounds. Methylene dichloride is the preferred solvent for film casting.

The resins are not made in the United Kingdom, possibly due, in part, to patent complications. They are expensive and the high price has restricted development; they have, however, found some applications in the electrical industry, for the manufacture of high quality protective helmets and for specialized components in the motor industry. The resins are frequently used unfilled but a modified material compounded with glass fibre, which shows a major increase in tensile strength, has been put on the market.

POLYFORMALDEHYDE

Polymers based on formaldehyde are generally referred to as polyacetals but are not to be confused with the polyvinyl acetals described on pp. 223 and pp. 224; they are, in effect, polyoxymethylenes of the general formula $(CH_2O)_n$ with modifications as described below. Although polymers of formaldehyde were first observed over a century ago, it is only within the last 10 years that it has been possible to produce high polymers of commercial value. These products are manufactured in the United States by two important companies, the DuPont group, under the trade name "Delrin" and the Celanese Corporation with the trade name "Celcon".

"Delrin" is made from a very pure formaldehyde, produced by decomposition of an alkali-precipitated low polymer; the polymerization is carried out in dry hexane using triphenyl phosphine as catalyst. Diphenyl methane, as a polymer stabilizer, and a trace of methanol, which acts as a molecular weight regulator, are added. The polymer, which is insoluble in hexane, is filtered off, washed with heptane and acetone and dried in vacuum.

The Celanese Corporation's product, "Celcon", is believed to be made by copolymerizing the formaldehyde trimer, trioxane, with about 2% of ethylene oxide; the main advantage of "Celcon" is said to be that it can be processed at rather lower temperatures than a pure formaldehyde polymer and there is, therefore, less degradation during the processing operations.

The material is not yet made in the United Kingdom but B.I.P. Chemicals Ltd. at Oldbury, near Birmingham, have announced that they are planning a plant to produce a poly-formaldehyde, also starting with trioxane.*

Processing and Properties

Polyformaldehyde can be processed by extrusion or injection moulding at temperatures nearing 200°C; "Delrin" especially, melts quite sharply at 180°C.

Density varies between 1·41 and 1·42 for various grades. "Delrin" has a tensile strength of about 10,000 lb/in^2 compared with "Celcon" at about 8000 lb/in^2, but the latter has a slightly lower water absorption. All polyformaldehydes have a very high impact strength which, even at temperatures as low as 0°C, is about 6 ft-lb/in. notch and rises to 7·8 ft-lb at 23°C; their stress crack resistance is also particularly good. The dielectric constant is 3·7–3·9 and is almost independent of temperature, while they have good electrical resistivity. Resistance to alcohols and aromatic hydrocarbons is excellent and the polymers have considerable resistance to other organic solvents but are readily decomposed by inorganic acids and alkalis. The maximum service temperature does not exceed 70–75°C and stabilizers are normally required to give good weathering properties.

Uses

The very high tensile strength and resistance to stress cracking make polyformaldehyde an excellent material for engineering applications and it is in this field that it is mainly used. Typical uses are for bushes, bearings, gears, instruments of all kinds, business machinery, telephone parts, car instrument housings and

*Plan now abandoned (Jan. 1970).

pump parts in contact with coolants and lubricants. Consumption in the United Kingdom has, so far, been very small as all the material has to be imported and, compared with other plastics, the price is very high.

POLYTETRAFLUOROETHYLENE

This polymer, often known as PTFE, was discovered just before the Second World War but commercial production did not start until 1950. It is made in the United Kingdom by I.C.I., who use the trade name "Teflon", but only to the extent of some hundreds of tons annually.

The monomer, tetrafluoroethylene, is an odourless gas which liquefies at $-76°C$; it is produced from chloroform and hydrogen fluoride by the following steps:

$$CHCl_3 + 2HF \longrightarrow CHClF_2 + 2HCl$$

$$2CHClF_2 \xrightarrow{\text{Pyrolysis}} CF_2{=}CF_2 + 2HCl$$

Polymerization is probably carried out in suspension under a pressure of 1000 lb/in^2 or more at a temperature of about 100°C: a peroxide is a suitable initiator. The polymer may be produced in the form of granules, as a fine powder or as an aqueous dispersion; the two former grades are used for moulding and extrusion, the latter for coating and impregnating.

Properties

PTFE has properties that are not matched by any other polymer in commercial production and these are summarized below:

1. Its heat distortion temperature is 130°C but it does not soften until 320°C; it never really melts but slowly evaporates and decomposes above 400°C.

2. No known organic solvents or plasticizers will attack it and,

even at high temperatures, the only inorganic materials that will react with it are molten alkali metals and elementary fluorine.
3. It has a density of 2·2, the highest of all commercial plastics, and its water absorption is nil.
4. Its electrical properties surpass those of polyethylene and polystyrene.
5. It is slightly flexible, waxy in feel and has a tensile strength of about 4000 lb/in².
6. Because of its chemical inertness it is difficult to make the polymer stick to other materials and its coefficient of friction is almost zero.

Processing

Its high softening temperature and peculiar behaviour at high temperatures make PTFE impossible to process by conventional techniques and methods resembling those of powder metallurgy are used. Solid components can be made by compressing cold granule polymer to the desired shape and then sintering at 370–390°C. Rods and tubes are made by compressing the cold powder and then forcing the compacted mass slowly through a heated die where, again, it is sintered. Rates of production are low because of the low rate of heat transfer.

Aqueous dispersions, suitably stabilized, can be spread out on smooth surfaces and converted by heat to thin films which have high chemical, heat and moisture resistance. Because of the anti-adhesive properties of the polymer the surfaces must be specially primed and treated to enable the films to adhere.

Uses

PTFE is an ideal material for engineering applications. Its high resistance to heat and chemicals make it excellent for gaskets and sealing rings under severe service conditions while its anti-sticking characteristics lead to its use for unlubricated bearings and as a coating for all kinds of plant handling sticky materials. A well-known domestic application is for coating the insides of saucepans and frying pans to prevent milk products, eggs and similar foodstuffs sticking to them. Very thin sheets have been used to replace mica in the development of compact high-frequency equipment.

Copolymers of tetrafluoroethylene and hexafluoropropylene are sometimes used instead of PTFE as they can be processed more easily.

OTHER FLUORINE-CONTAINING POLYMERS

The unique properties of PTFE have naturally led to the investigation of other polymerizable fluorine compounds, such as vinyl fluoride and vinylidene fluoride, but it has on the whole been found that any improvement in properties is not worth the greatly increased cost. Probably the most important compound is chlorotrifluoroethylene which may be polymerized with benzoyl peroxide as an initiator and chloroform as a chain stopper to limit the molecular weight attained. The polymers are less resistant to heat and chemical attack than PTFE but are more easily processable on standard equipment; their softening points range from 175° to 300°C according to molecular weight.

COUMARONE–INDENE RESINS

Coumarone and indene together make up to about 30% by weight of the heavy naphtha fraction of coal tar boiling between 160° and 190°C. Coumarone boils at 172°C and indene at 180°C. They are not normally separated but are copolymerized *in situ* using the other hydrocarbons present as a solvent. The usual catalyst is dilute sulphuric acid but Friedel–Crafts catalysts or phosphoric acids are also effective. A chain copolymer is formed in which the arrangement of the coumarone and indene molecules is believed to be completely random: it may be represented as follows:

Polymerization with sulphuric acid is very rapid and, after separation of any tar formed, the acid is neutralized, the aqueous layer separated and the solvent hydrocarbons steam-distilled off. Any low molecular weight polymers are removed at the same time and the steam distillation is continued until the desired softening point is reached. Molecular weights are low, ranging from 1000 to 3000.

The resins can vary from viscous liquids to friable solids melting as high as 150°C. They are thermoplastic, have excellent electrical properties and do not readily oxidize; they are soluble in hydrocarbon oils and many vegetable oils and are used in varnishes. Their main use, however, is in flooring and as a processing aid in the compounding of rubber. Addition of coumarone–indene resin actually improves the strength of a mineral-filled synthetic rubber.

SILICONES

Silicon, like carbon, is tetravalent and is capable of forming compounds with the same structure as the corresponding carbon compounds. The hydride, silane, SiH_4, is analogous to methane and may be regarded as the parent compound of the silicon polymers in commercial use.

Most of the basic work on organic derivatives of silicon was carried out by F. S. Kipping over the astonishingly long period from 1899 to 1944. The actual development of polymers resulted from an attempt to synthesize a product which would combine the properties of glass with those of transparent organic polymers and took place in the laboratories of the Corning Glass Works in the United States. In 1942 a partnership between the Corning

Glass Works and the Dow Chemical Company was set up to develop silicone polymers commercially and operated as the Dow Corning Company. Later, General Electric in the United States also became involved in work in this field. In the United Kingdom, Albright and Wilson became the selling agents for Dow Corning silicones and, when the market had sufficiently developed, a joint company, Midland Silicones Ltd., was set up in 1954 to manufacture silicones near Barry in South Wales. In 1955 I.C.I. also began manufacture at Ardeer under licence from General Electric.

Due to the large number of different silicone products made and to the fact that many of the intermediates can be made in multi-purpose plant, capacities are somewhat indefinite but it is believed that Midland Silicones can manufacture 4000–5000 t/a and I.C.I. have new plant just coming on stream which will raise their total capacity to more than 5000 t/a. Current consumption in the United Kingdom is believed to be about 3000 t/a and is growing at a rate of 10–15% per annum. A large proportion of both companies' output is exported, including many of I.C.I.'s intermediates to Rozenberg in Holland, where they have a silicones unit.

Chemistry of Silicones

Silicone polymers are produced by the hydrolysis of methylchlorosilanes; the hydrolysed compounds are known as polysiloxanes. Dimethyldichlorosilane acts as a difunctional compound and gives thermoplastic polymers while the trifunctional methyltrichlorosilane will yield thermosetting polymers.

$$
\begin{array}{c}
CH_3 \\
| \\
Cl-Si-Cl + H_2O \longrightarrow \\
| \\
CH_3
\end{array}
\quad
\begin{array}{c}
CH_3 \quad CH_3 \\
|\qquad\quad| \\
-O-Si-O-Si-O- \\
|\qquad\quad| \\
CH_3 \quad CH_3
\end{array}
$$

Dimethyldichlorosilane

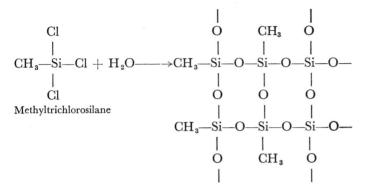

Manufacture of Chlorosilanes

The process most commonly used for commercial production is to pass an alkyl chloride, for example methyl chloride, through a mixture of elementary silicon with 10% of a copper catalyst at 250–80°C. The issuing gases consist of a mixture of methyl chlorosilanes and unreacted methyl chloride; it is condensed and separated into its constituents by fractional distillation. The silanes consist of about 73% of the dimethyl compound, 10% of the monomethyl derivative and 6% of the trimethyl with some by-products. Other alkylchlorosilanes are made by an analogous process.

Phenylchlorosilanes, required for some types of silicone rubber, are made by a similar process but the copper is replaced by silver as a catalyst and a higher reaction temperature of about 400°C is needed. The principal product is diphenyldichlorosilane. All the organochlorosilanes are very reactive and flammable and must be stored in an inert atmosphere under completely dry conditions.

SILICONE POLYMERS (POLYSILOXANES)

There are three basic types of product made commercially:

1. *Silicone fluids and greases:* these are usually low molecular

weight polymers made from dimethyldichlorosilane or, sometimes, diphenyldichlorosilane.

2. *Silicone resins:* these are thermosetting materials made from dimethyldichlorosilane with addition of varying quantities of monomethyl trichlorosilane.

3. *Silicone elastomers:* these are normally high molecular weight polymers made from dimethyldichlorosilane and/or diphenyldichlorosilane.

United Kingdom output is believed to consist of about two-thirds of group 1 and one-sixth each of groups 2 and 3.

Silicone Fluids and Greases

Dimethyldichlorosilane, often blended with a little trimethyl-monochlorosilane to act as a chain stopper to keep the molecular weight low, is hydrolysed by water or dilute hydrochloric acid. The reaction is virtually instantaneous and is exothermic so that cooling is necessary. An oily fluid separates which, at this stage, is a mixture of cyclic compounds and diols. The oil is run off and heated with a strong acid or base, which breaks down the cyclic polymers, further polymerizes some of the shorter chains and depolymerizes the longer ones; the process is known as equilibration and the whole effect is to make a product with a more homogeneous molecular structure.

The silicone fluids so obtained are colourless, odourless, non-volatile and non-toxic; they are soluble in hydrocarbons but insoluble in oxygenated solvents and water and are not attacked by any chemicals except concentrated mineral acids. The fluids are stable up to 150°C in air and to much higher temperatures in the presence of an antioxidant, such as a substituted phenol or amine, but are flammable if external heat is applied for a long time, due to molecular breakdown leading to low molecular weight flammable compounds.

The chief applications of silicone fluids stem from their very high resistance to mechanical shear and from their water repellent and anti-adhesive properties. Added to wax polishes they permit a very thin film of wax to be obtained with a minimum of rubbing while small amounts are added to paints to improve the gloss and reduce flaws in the film. An aqueous emulsion of the fluids can be applied to textiles to give an excellent water repellent finish, while leather can be made similarly water repellent by immersion in a 15% solution of a silicone fluid in a hydrocarbon solvent. Applied to paper, silicones give both water repellency and anti-adhesive properties; they may also be used as mould release agents.

By blending the fluids with fillers, such as lithium soap and carbon black, excellent lubricants and greases are obtained which are effective between subzero temperatures and 200°C; by thickening with alkyl urea compounds, greases effective at still higher temperatures can be made. Replacement of the methylchlorosilanes by phenyl compounds also increases the upper working temperature of lubricants, but at the expense of the viscosity temperature coefficient. These lubricants are, of course, expensive but have many small scale applications, such as lubricating delicate plastic gears and bearings in synchronous clock motors, for many kinds of meter and for the operating mechanism of valves working in a wide range of liquids and gases. Typical high temperature applications are for oven doors and for conveyors passing through industrial ovens.

Silicone resins

Production methods for silicone resins vary in detail but depend, in principle, on hydrolysis of a dialkyl chlorosilane containing some monoalkyltrichlorosilane. A general method is to mix the silanes with a slight excess of water and heat, with stirring, until polymerization has reached the desired stage, as shown by viscosity measurements. Alternatively, a low molecular weight polymer may first be formed, dissolved in a high boiling hydrocarbon or chlorinated hydrocarbon solvent and further polymerized by heating, often in the presence of a catalyst, such as an alkylamine or a methyl naphthenate.

Silicone resins can be cured by simple heating. They may be applied as coatings where temperature resistance and water repellancy are required. A typical application is for coating the tins used for baking bread, where they supersede greasing and make easy release of the bread possible; in the same way they can be applied to moulds used in the plastics and rubber industries where sticking presents a problem.

Silicone resins can be used for laminating and have obvious outlets when temperature-resistant fillers, such as glass or asbestos, are used. The mechanical properties of such laminates are inferior to the corresponding materials made from phenolic resins, but they do not deteriorate until the working temperature approaches 400°C. Another major outlet for the resins, which depends on their resistance to high temperature, is for electrical insulation where they permit the use of higher currents in a given conductor than are possible with other insulants. This makes possible the production of small size, but high capacity electric motors.

At a frequency of 60 c/s the resins have a dielectric constant of about $3 \cdot 5$ and a power factor at normal temperature of about $0 \cdot 01$.

Silicone Elastomers

Elastomers are made by hydrolysing very pure dimethyl- or diphenyl-dichlorosilane to give high molecular weight thermoplastic products. Equilibration is necessary to break down ring compounds and to homogenize the polymer as far as possible; ferric chloride, sulphuric acid or caustic soda are the reagents used.

Fillers and curing agents are then incorporated on mixing rolls by conventional rubber processing techniques. Most rubber fillers can be used, with the exception of carbon black which interferes with the cure and, in addition, has limited heat stability. The best reinforcing filler is a very finely divided silica, obtained by burning silicon tetrachloride and hydrogen. This can give tensile strengths up to 2000 lb/in² and elongations up to 600%, although average samples only attain about half these values. Peroxides, such as benzoyl peroxide, are used as curing agents; they are believed to act by producing free radicals by oxidation of CH_3 groups so that activated CH_2 groups on adjoining chains can link up. The uncured material can be extruded or moulded to the required shape and then cured under some pressure for 15 min at 120°C. This gives the required "set" but, for optimum properties, the article must be baked at 250°C for several hours.

The most important property of silicone elastomers is that they retain their properties over a very wide temperature range, for example from −80° to 200°C, and are stable even at 300°C for a considerable period of time. For maximum low temperature effectiveness a methyl phenyl polysiloxane gives the best results. The rubbers show great bounce and flexibility but have not the tensile strength or abrasion resistance of conventional rubbers. They have been used for the production of gaskets and seals for use under high temperature conditions. Glass cloth coated with silicone elastomer is used as ducting for the anti-icing and heating systems of modern aircraft. The rubbers are also used in surgery, for example to make artificial heart valves.

While these elastomers are resistant to lubricating oils, they break down in the presence of hydrocarbon solvents; this defect can be overcome by hydrolysing alkylsilanes containing some fluorine substituents in the alkyl groups. Many curing systems are being investigated and there is much activity in the development of new elastomers having special properties.

The well-known bouncing putty is made by heating the thermoplastic polymer with 5% of its weight of boric acid. It can be moulded in the hand like putty, flows under its own weight, can be shattered by a sudden blow and yet will bounce when dropped on a hard surface.

READING LIST

Polyamide Resins, by Floyd, Reinhold, New York, 1958.
Nylon, by B. W. Hirsh, Plastics Institute Monograph.
Vinyl and Related Polymers, by Schildknecht, John Wiley, New York, 1962.

Acrylic Esters, by Horn, Reinhold, New York, 1960.
The Chemistry and Physics of Polycarbonates, by H. Schnell, Interscience, New York, 1964.
Polyaldehydes, by G. Vogl, Arnold, London, 1967.
Fluorocarbons, by Rudner, Reinhold, New York, 1958.
Silicones, by Freeman, Iliffe Press, London for the Plastics Institute, 1962.
Silicones, by Meales and Lewis, Reinhold, New York, 1959.

PART IV

Natural Polymers

CHAPTER 12

Cellulose and Casein

THE two naturally occurring polymers, cellulose and casein, were the basis of the first commercial plastics materials. They are no longer of major importance; the use of casein has declined greatly while the use of cellulose plastics, though still substantial, tends to be confined to specialized outlets.

CELLULOSE PLASTICS

Cellulose occurs widely in Nature and has been called the chemical that grows; it is present in all plant structures but the sources of all commercial cellulose are cotton and wood. Cotton is almost pure cellulose and the short fibres (linters) which remain after the ginning process are an important source of the polymer for plastics use. Wood consists essentially of cellulose fibres bound together by another natural polymer, lignin, and the two materials are separated in the preparation of wood pulp, which consists mainly of cellulose, as the first step in paper manufacture. Purified cellulose for plastics manufacture is produced from cotton linters or from wood pulp by treatment with caustic soda and bleaching.

Cellulose Chemistry

The stereo-chemistry of cellulose is quite complicated but it can be regarded simply as a carbohydrate with the empirical formula $C_6H_{10}O_5$. The repeating unit in its structure may be written:

$$\begin{array}{c}\text{CH}_2\text{OH}\\ |\\ \text{C}\!-\!\text{O}\quad\text{H}\\ |\quad\quad\backslash\\ \text{H}\quad\quad\text{C}\!-\!\\ -\text{O}\!-\!\text{C}\quad|\quad\quad|\\ |\quad\text{H}\quad\text{H}\quad\text{C}\!-\!\\ \text{H}\quad\text{C}\!-\!\text{C}\\ |\quad|\\ \text{OH}\quad\text{OH}\end{array}$$

It will be seen that the molecule has three hydroxyl groups which may be esterified or etherified. Derivatives produced in this way form the raw material of the plastics applications to be described.

Regenerated Cellulose

Cellulose, in its natural state, is a white fibrous material and is not a thermoplastic; it is only by production of derivatives, as described in the previous paragraph, that materials with thermoplastic properties may be obtained. Cellulose may, however, be treated with caustic soda and carbon disulphide to form a xanthate which may then be decomposed to regenerate the cellulose as a transparent fibre or film. Fibres produced in this way are used on a large scale for textiles, described originally as "artificial silk" and now generally known as viscose rayon. The films are transparent and are very similar in appearance to film made from polyolefins or PVC. Although permeable to water vapour in their raw state, they may be coated in various ways to make them vapour proof and give them heat sealing properties. They are widely used as packaging materials and are known generally as "Cellophane", although, strictly speaking, this is a trade mark. The largest producer in the United Kingdom is British Cellophane Ltd., a subsidiary of Courtaulds, with an output believed to be between 40,000 and 50,000 t/a.

Cellulose Nitrate

Cellulose nitrate compounded with camphor is the basis of the first commercial synthetic plastic invented by Parkes in 1862 (see p. 7). The Parkesine company, formed to exploit the invention, was not very successful and was wound up after 10 years or so. Manufacture of the material was continued, however, by Spill, a former associate of Parkes, and was placed on a sounder commercial basis with the formation in 1877 of the British Xylonite Company, now part of the Bakelite Xylonite Group. Development proceeded more rapidly in the United States where the name "Celluloid" originated.

Although the material has many excellent properties, its high flammability has always been a disadvantage and it has gradually been displaced by the newer, less flammable plastics which are now available. Over 10 thousand tons a year are still made around the world by the original process, which is described below.

Manufacture of Cellulose Nitrate

Cotton linters, purified by treatment with caustic soda and bleaching, are dried to a water content of less than 2% and then treated in batches with a mixture of strong sulphuric and nitric acids in a ratio of about $2:1$. Care must be taken not to let the reaction get out of control. After a period of digestion, which may be longer if a lacquer is to be made than if a celluloid is to be produced, the residual acid is separated in a centrifuge and the nitrated cellulose washed repeatedly with water to remove the last traces of acid. The bulk of the water is removed centrifugally and the last traces are eliminated by washing with ethyl, isopropyl or butyl alcohol. The nitrated fibres are similar in appearance to the original material and should contain about 11% of nitrogen. Higher nitrogen contents are required for explosive grade nitrocellulose and more concentrated acids are used.

Conversion to Celluloid

Celluloid is cellulose nitrate plasticized with about 30% of camphor; sometimes other plasticizers, such as phthalates, are used as well and pigments can be added to give the desired colour. Camphor has been used since Parkes's time, no better plasticizer having been found. It has to be incorporated with the help of alcohol, which is not itself compatible with cellulose nitrate but becomes so in the presence of camphor. The ingredients are kneaded in a suitable mixer, when the cellulose nitrate loses its fibrous structure and a homogeneous mix is obtained. This operation takes about an hour and its end is judged by the appearance of the material, which is then a soft dough. The product is then filtered to remove any extraneous impurities by being forced under high pressure through a sheet of calico, backed by strong brass gauze and supported on a heavy brass plate from which the material emerges in the form of weak filaments. Much of the alcohol is then removed in a further mixing on hot rolls and the material is sheeted off. The process is illustrated in Plate X. These thick sheets, about 4 ft 6 in. × 2 ft, are then compressed into a solid block 4–6 in. thick at a carefully controlled temperature followed by cooling under pressure, the whole process taking about 5 hr.

The blocks are then sliced on a moving table to which the block is fixed. The table passes backwards and forwards under a knife inclined at an angle of 30° to the block and, by controlling its setting, sheets of any desired thickness between 0·005 in. and 1 in. can be obtained. The material in the sheets still contains 10–12% of alcohol which is removed by hanging them in a warm air stream at 30°C in solvent recovery stoves for 24 hr; drying is then completed at 50°C for a time varying from 3 to 56 days according to the thickness of the sheet.

Finally, any slight surface irregularities are removed in a polishing process in which the sheets are heated under hydraulic pressure between highly polished metal plates. Celluloid can also be formed into rods, tubes and profile shapes by extrusion of a composition containing a little alcohol.

The rather complicated solvent-assisted celluloid process is unique in allowing the production of many attractive three-dimensional colour effects, such as tortoiseshell, derived from the

PLATE X. Milling celluloid on hot rolls. (By courtesy of Bakelite Xylonite Ltd.)

mechanical manipulation and juxtaposition of two or more differently coloured materials in successive pressing and slicing operations.

Cast film is made from cellulose nitrate by mixing with 10–15% of camphor and a 70–30 mixture of alcohol and ether or acetone until a smooth viscous "dope" is obtained. The dope is filtered and allowed to fall through a narrow horizontal slit on to a moving endless band, contained in an enclosed space, heated sufficiently to drive off most of the solvent while the band performs almost a complete turn; just before reaching the casting point again the film is stripped off and, still under tension, is further seasoned and the removal of solvent completed by being passed over a series of rollers, contained in a controlled atmosphere cabinet and maintained at gradually increasing temperatures up to about 40°C. A final set of chilled rollers cools the film before it is wound on to a roll; a typical celluloid film is 0·003 in. thick.

Properties and Applications

Uncoloured celluloid is a nearly transparent, almost water-white solid, softening below 100°C and considered unsuitable for use much above room temperature. It has a tensile strength of about 5000 lb/in², an impact strength of about 2·8 ft-lb/in. of notch and a compressive strength of some 20,000 lb/in². The sheet can readily be moulded when heated to its softening point and moderate pressure in a simple mould is satisfactory. It can also be blow moulded and stands up well to deep drawing.

At one time it was the only transparent thermoplastic material available and had a wide range of applications, especially as cinematograph film; its flammability has, however, always been a serious disadvantage and cine film is now made from cellulose acetate, described in the next section.

It has good chemical resistance and is sometimes used for battery cases. Celluloid tubes can be shrunk on to metal tubes and adhere firmly so that, for example, corrosion-resistant bicycle handlebars can be made in this way. Other applications are the production of knife handles, combs, spectacle frames and table tennis balls, for which celluloid is still unsurpassed. Its resistance to ultraviolet light is rather poor so it cannot be used in applications involving extensive exposure to direct sunlight.

Cellulose Acetate

Cellulose acetate was known in the nineteenth century and many workers experimented with it in an attempt to replace the highly flammable celluloid. All attempts to produce a satisfactory thermoplastic material failed, however, until after the First World War when plasticizers became available. Cellulose acetate dopes for aircraft fabric were made in 1914–18 by Dreyfus at Spondon, near Derby and, from this operation, British Celanese Ltd. developed. Cellulose acetate fibre (acetate rayon) production was started in the early twenties but satisfactory plastics were first made in 1927, when organic phosphates were found to have good plasticizing properties. In the 1930's cellulose acetate was the material which enabled injection moulding to be developed but, after the Second World War, competition from polystyrene halted its growth. It has, however, maintained certain substantial uses but no large growth can be expected; total production in all forms, except fibres, is currently in the neighbourhood of 10,000 t/a, of which 4500 tons are moulding powder, the remainder being sheet, film and extruded products.

Manufacture

The acetylation of cellulose has been widely studied, mainly with the object of producing improved acetate rayon. Cotton linters are again the main raw material and are given a pretreatment with dilute caustic soda solution, bleached and dried. The treated linters are then suspended in acetic acid and treated with three parts by weight of acetic anhydride to one part of cellulose, with a trace of sulphuric acid as catalyst. The reaction is started at 0–5°C, with stirring, until a thick slurry is formed when acetylation proper begins and the temperature is allowed to rise to 35–40°C; the slurry gradually loses its fibrous appearance and its viscosity falls as acetylation proceeds. After the main acetylation reaction is over the temperature begins to fall; samples are examined periodically under a microscope and the reaction stopped when all the fibres have disappeared. The product is now in the form of the triacetate.

A high degree of acetylation decreases the water absorption of the plastic and increases its dimensional stability but has an unfavourable effect on flow temperature and compatibility with plasticizers. However, solvents and techniques have been developed for converting triacetate into film as well as fibres. For other plastics and dopes a product between the diacetate and triacetate is required and this is produced from the crude reaction product by a "ripening" or saponification process. The reaction mixture is blended with 20–25 parts of water per 100 of cellulose, cooled if necessary to 20°C and allowed to stand for about 3 days; progress is assessed by acetone solubility tests and, when complete solubility is attained, saponification is stopped by the addition of sodium acetate. Cellulose acetate flake suitable for producing moulding powders is precipitated by agitating the ripened mixture with water containing sufficient alkali to neutralize the sulphuric acid catalyst present; the dilute alkali solution is added slowly to avoid the formation of lumps and the precipitated mass is allowed to stand for 15–20 min to harden before being separated by centrifuge. The centrifuged material is washed to free it from acetic acid but still contains sulphur as the acetosulphate; this is removed by heating the impure cellulose acetate to 100°C with a trace of acetic acid, which breaks up the sulphur complex without damaging the acetate proper. The two acids are then washed out and the cellulose acetate dried at temperatures not exceeding 100°C until the water content is about 2–3%. This is the material used for the production of moulding powder and film; if the film is for photographic use the acetate is subjected to a special treatment for the removal of metallic impurities.

In a newer alternative process, operated under pressure, methylene chloride is used as a solvent instead of acetic acid; it makes temperature control during acetylation easier and is distilled off at the end by reducing the pressure, which brings about precipitation. Substantial quantities of dilute acetic acid are produced as a by-product in both processes and the whole operation only becomes economic if this can be reconverted to acetic anhydride or used elsewhere. An important advantage of the methylene chloride process is the smaller quantity of dilute acetic acid needing recovery.

Production of Moulding, Extrusion and Casting Compounds

Unplasticized cellulose acetate softens not far below its decomposition temperature and, in addition, does not flow readily under heat and pressure. The plasticizers normally used are high boiling esters; dimethyl or diethyl phthalate, with a little acetone to assist compatibility, is particularly suitable; 25–40 parts of plasticizer are added to 100 parts of flake. The plasticizer must have a high degree of compatibility with the cellulose acetate, be reasonably non-volatile, not readily extracted, stable at processing temperatures, stable to ultraviolet light and, for products which come in contact with foodstuffs, it must also be tasteless and non-toxic.

The plasticized flake can be slightly off-white so a trace of blue dye is added when a colourless product is required; for coloured products suitable pigments may be added but must conform to the same stability and toxicity criteria as plasticizers.

Fillers such as mica, carbon black, zinc oxide or wood flour are added to increase hardness, modulus of elasticity and flexural strength, but they impair the appearance of the finished material. If the finished product is to be used in sunlight a small quantity of stabilizer such as hydroquinone, a benzyl ether or various amines is added.

The flake is converted to moulding powder on conventional equipment; the preliminary mixing is carried out in a Banbury type mixer and the mixture thus formed is fluxed on heated rolls. The cooled product from the rolls is ground and forms the finished moulding powder.

Cellulose Acetate Sheet and Film

Sheet is normally produced by a process similar to that already described for celluloid on p. 248. Plasticizers, such as dimethyl phthalate and triphenyl phosphate, are used in the preliminary mix together with solvents such as acetone and benzene to assist processability. The other steps of filtration, sheeting, solvent removal, lamination and, finally, slicing follow. Sheet can also be made by extruding compounds without solvent through a slot die, using temperatures much higher than are used in the celluloid process.

Cellulose acetate film is also cast by a method similar to that used for celluloid film. Triacetate has, in recent years, been the only material used for photographic film but there are now signs that Terylene film is beginning to penetrate this market. When photographic film is being made an exceptionally high standard of cleanliness is essential and the whole operation is carried out in enclosed equipment.

Properties and Application of Cellulose Acetate

Typical properties for cellulose acetate are: density $1 \cdot 23$–$1 \cdot 33$, tensile strength 3000–5500 lb/in^2 and impact strength about $3 \cdot 3$ ft-lb/in. of notch. It can be injection moulded or extruded; the processing temperatures vary quite widely between 130° and 240°C while the mould shrinkage is $0 \cdot 3$–$0 \cdot 6 \%$. It is not so tough or water resistant as celluloid but is much less flammable; dimensional stability is rather poor, due to the relatively high water absorption, but electrical properties are fair. The finished material is resistant to aliphatic hydrocarbons, oils and greases, but has relatively poor resistance to other organic solvents, acids and alkalis.

Although the newer thermoplastics have taken much of the market, cellulose acetate moulding powders are still used for injection moulding and, especially where tough products are wanted in attractive, clear colours, they have found many uses in decorative applications. Toys, buttons, shoe heels, door handles, combs, spectacle frames, car steering wheel covers and umbrella handles are typical of the products moulded from this material. All National Health and many other spectacle frames are cut from sheet, which also finds applications for toys, lampshades and similar articles.

Translucent cellulose acetate sheet is widely used for attractive clear display cases and covers and for giving a glossy finish to articles such as gramophone record sleeves, although polypropylene is largely replacing it in the latter application.

Cellulose Acetate-Butyrate

This product, commonly called CAB in the industry, is not made in the United Kingdom but is imported from the United States and Germany. It is manufactured by treating cotton linters, first with 40–50% sulphuric acid for 12 hr and then with 100% acetic acid for a further 12 hr; the next step is esterification at 35–50°C with a mixture of 1 part of butyric acid to 2 parts of acetic anhydride and a trace of sulphuric acid as catalyst; the product is then finished like cellulose acetate.

CAB flake has an advantage over the straight acetate in that it is fully compatible with a much wider range of plasticizers and the finished products are more moisture and heat resistant, have better dimensional stability and stand up better to weathering; they are not, however, resistant to aromatics or other organic solvents. Compounding and manufacture of moulding powders is similar to that for cellulose acetate; the product is particularly good for injection moulding; its main disadvantage is that it costs more than the straight acetate and its price advantage over other plastics is marginal. It is used for tool handles, thermoformed signs, typewriter keys and similar applications;

extruded with integral bright metal strip it is excellent for decorative trim.

Cellulose Ethers

As noted on p. 246, it is possible to etherify the hydroxyl groups in cellulose; the products are not normally regarded as plastics but have uses on the fringe of the plastics industry. For the sake of completeness they are briefly described below.

Methyl Cellulose

Alkali cellulose, made by masticating wood pulp with concentrated caustic soda solution, is reacted with methyl chloride; the methyl cellulose formed is washed with very hot water and separated by centrifuging; an average of about $1 \cdot 8$ of the three hydroxyl groups is etherified. It is used as a thickening agent in cosmetic preparations and in textile printing pastes, as an adhesive in the leather industry and in paint remover formulations.

Ethyl Cellulose

This is made in a similar way to methyl cellulose, substituting ethyl chloride for methyl chloride. Materials of low ethyl content are water soluble but, with ethyl contents of $2 \cdot 4$–$2 \cdot 5$ per cellulose unit, the products are insoluble in water but soluble in most organic solvents except aliphatic hydrocarbons.

Ethyl cellulose of the latter type is tough and does not become brittle, even at $-40°C$; it is compatible with a wide range of plasticizers, resins and waxes and may be used to raise the melting point of low melting resins and to improve their hardness and toughness.

Hydroxy Ethyl Cellulose

Prepared by the action of ethylene oxide on alkali cellulose, this product is insoluble in water but soluble in caustic soda; it is used as a sizing agent in the paper and textile industries. A modified form, of low ethylene oxide content, is water soluble and has applications as a thickening agent and adhesive.

Sodium Carboxy Methyl Cellulose

This is prepared by reacting alkali cellulose with sodium monochloroacetate. Products with a high degree of substitution are freely soluble in water and are used in wallpaper adhesives and in detergent formulations.

CASEIN PLASTICS

These are long-established products, introduced in Germany as long ago as 1900 and, at one time, reaching an output of several thousand tons annually in this country; their importance has declined and production is now small. The materials were made in the United Kingdom by Erinoid Ltd., now part of B.P. Chemicals (U.K.) Ltd. and some other companies, and are interesting as one of the few examples of a protein plastic that has attained commercial production.

The casein is prepared from fresh skimmed milk, heated to 40°C and mixed with a small quantity of an enzyme; after a period of standing the curd formed is broken up and the casein allowed to settle. The liquid whey is decanted and the solid material washed repeatedly with water and dried at 65°C.

Casein plastics are manufactured by mixing the solid casein with about 30% of its weight of water, together with dyes, pigments, etc., and extruding the mixture in the form of rods or tubes. Sheets are formed by flattening rods in a press. The products are machined to size if required and then hardened (cured) by immersion in 4–5% formaldehyde solution for several weeks.

The chemistry of the process is extremely complicated and will not be dealt with. The final product has a density of about 1·3 and may be polished by soaking in a solution of sodium hypochlorite; it is almost non-flammable and can be machined just like wood. It has, however, a poor resistance to water, it discolours at 70°C and chars readily. It has been widely used for making buttons and for such products as buckles, knife handles, brush backs, cigarette holders, umbrella handles and similar articles.

OTHER NATURALLY OCCURRING POLYMERS

Many polymeric materials are found in Nature and, although none of them has major outlets as a plastic, some of them have small scale applications as plastics and, in other ways, on the fringe of the industry. They do not merit a chapter to themselves and a brief description is given below.

Shellac

Shellac is one of the earliest known thermoplastics and has been known in India, its main country of origin, and other countries of the Far East for many thousands of years. The resin is the protective coating formed by the lac insect, which lives on certain species of trees in India. The raw resin is scraped from the twigs and is refined by washing and bleaching processes; the refined resin is soluble in alcohol and its solutions form the basis for excellent wood lacquers giving films of great brilliance and high gloss.

Shellac softens at about 80°C and has been used as a thermoplastic, when compounded with fillers such as powdered mica, for production of electrical insulators and coil formers. Its major plastic use for many years, however, was for the production of gramophone records but it has now been almost entirely replaced in this application by PVC copolymers.

Rosin

Rosin is a constituent of the resin exuded by pine trees and consists mainly of the anhydride of abietic acid. It is used to modify phenol–formaldehyde resins (see p. 78) to make them oil-soluble and has its main outlets in the surface coating, paper and detergent industries.

Bitumen

Bitumens are, strictly speaking, naturally occurring mixtures of high molecular weight, complex hydrocarbons but the term is often loosely applied to the heavy residues from petroleum stills and similar materials. When loaded heavily with mineral matter they are used in vast quantities for road surfacing and, when compounded in various ways, have many uses in manufacture of adhesives and waterproofing materials. Properly compounded they have good chemical resistance and are often used for moulding accumulator cases.

Lignin

Lignin has been mentioned on p. 245 and forms about 30% of the solid matter of wood; it is potentially available in very large quantities as a by-product in the manufacture of wood pulp. Limited uses have been found for it as a binder but it is of no importance in the plastics industry.

Starches

Starches are polymers which have many uses as adhesives, binders and sizes but have no uses as plastics.

READING LIST

Cellulosic Plastics, by Yarsley, Flavell, Adamson and Perkins, Iliffe Books for the Plastics Institute, London, 1964.

Cellulose Acetate Plastics, by Stannett, Temple Press, London, 1950.

Cellulose Acetate, by Courtaulds Ltd., Plastics Division, Little Heath, Coventry, 1950.

Casein Plastics and Allied Materials, by J. H. Collins, Plastics Institute, London, Monograph, 1952.

PART V

Elastomers

CHAPTER 13

Synthetic Elastomers

ELASTOMERS were mentioned in the introduction to this book without any attempt to define them precisely; the simplest definition is based on performance—an elastomer is a material which is capable of being stretched to at least double its original length and which returns rapidly to its original length when the stretching force is removed. Elastomer is a synthetic word devised fairly recently to cover a group of polymeric materials with rubber-like properties of which natural rubber was the first and is still the most important. Many of the newer synthetic rubbery materials differ so widely from the natural product, both in their chemical make-up and in their range of applications, that the generic term elastomer is thought to be preferable to the older description of them as synthetic rubbers.

Processing methods for rubber and thermoplastics are very similar, and rubber-processing machinery has often been modified to meet the needs of the plastics industry. Although it is not obvious from their appearance, elastomers may be regarded as intermediate between thermoplastic and thermosetting resins, with the polymer molecules forming a loose network structure containing a relatively small number of cross-links; it is this loose structure which gives the material its elastomeric properties. The formation of the cross-links usually takes place during "vulcanization"; it is basically similar to the curing of a thermoset and may, in many cases, be carried so far as to produce an infusible, insoluble material indistinguishable from a normal thermoset. Such a product from natural rubber is called ebonite.

Natural rubber was in use on a large scale before plastics became industrially important and, because of its unique properties, it soon became essential in a wide range of industries and especially, of course, in transport; U.K. consumption reached an all-time high in 1956 at 200,000 tons. The urge to produce a synthetic substitute was twofold; because the production of natural rubber is confined to tropical countries, supplies to the large industrialized communities of Europe and North America can be cut off in time of war. Rubber has become essential, both for modern war and for normal peace-time living, and the effects of shortage were felt severely by the Central European Powers during the First World War. These effects were felt even more severely in the United Kingdom and North America after the capture of Malaya by the Japanese early in the Second World War. It was these events that led to the massive production of synthetic general-purpose rubbers. In more recent years the amount of natural rubber available became quite insufficient to meet the rapidly growing demand and this has been an added reason for expanding the production of synthetic materials. A less compelling reason for finding a synthetic substitute was the limitation on use imposed by the properties of the natural product. This stimulated the search for special-purpose materials; this was partially successful in the early 1930's and has now produced synthetic elastomers that retain their properties over a very wide range of conditions.

Before proceeding to a detailed description of the synthetic materials a short account of the constitution of the natural polymer is desirable. Much academic work on this subject was carried out in the early years of this century and it was soon established that the monomer is 2-methyl-1,3-butadiene, commonly called isoprene. Syntheses of the monomer were devised but, although it could be polymerized, all attempts to produce a synthetic "natural" rubber failed. It was not until the early 1950's, following the work of Natta and others on stereospecific catalysts, that the natural product was duplicated synthetically.

STRUCTURE OF ISOPRENE POLYMERS

The simplest polymerization reaction of dienes is the formation of a 1,4-chain polymer with simultaneous transfer of the residual unsaturation to the 2,3-position giving, with butadiene, a chain of the type:

$$-CH_2CH=CHCH_2CH_2CH=CHCH_2CH_2CH=$$
$$CHCH_2CH_2CH=CHCH_2-$$

The presence of the residual double bond in the chain makes geometrical isomerism possible; thus isoprene can undergo a 1,4-polymerization to form *cis-* and *trans-*polymers as follows:

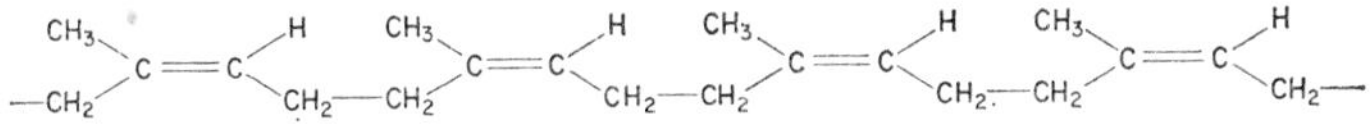

cis—Polyisoprene

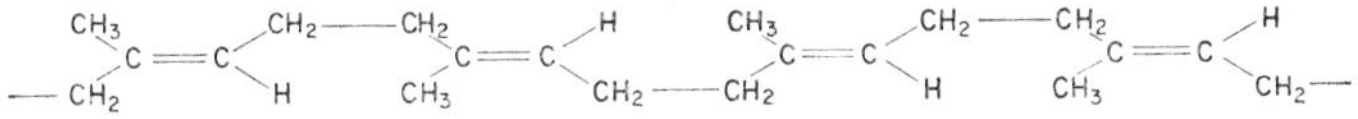

trans—Polyisoprene

Natural rubber is almost pure *cis*-polyisoprene while other natural rubbery polymers, such as gutta-percha and balata, consist mainly of *trans*-polyisoprene.

EARLY DEVELOPMENT OF SYNTHETIC RUBBERS

With the knowledge that rubber was basically polyisoprene it was to be expected that the first attempts to produce a synthetic would centre around polymerization of the parent hydrocarbon and closely allied compounds. The first synthetic rubber to approach commercial acceptance was produced in Germany during the 1914–18 war by polymerizing 2,3-dimethyl butadiene obtained from acetone via pinacone. The product had many defects, being especially poor in mechanical strength; although some 2500 tons were made in the period during and immediately after the war, as soon as natural rubber became available to German industry again, its production was abandoned.

The first commercially successful material was a special-purpose rubber produced and marketed by DuPont in the United States in 1931 under the trade name of "Neoprene". This is a polymer of 2-chlorobutadiene and is still produced commercially on a substantial scale. In the decade before the Second World War rubbery polymers and copolymers of iso-butene were discovered in Germany and the United States, and copolymers of butadiene with styrene and with acrylonitrile were brought into commercial scale production in Germany as "Buna-S" and "Buna-N" respectively. Much work on these copolymers was also done in the United States so that, when the fall of Malaya virtually cut off supplies of natural rubber, the production of styrene–butadiene rubber could be started on a very large scale almost immediately, albeit at very high cost.

The early styrene–butadiene rubbers had many disadvantages compared with the natural product. While these had to be accepted under war-time conditions, they stimulated improvements in the product which would have taken many years to bring about under more normal conditions. In consequence it was found that, by the time the natural material became freely available again, the synthetic product could compete with it on a price–performance basis in many applications and especially in car tyres, an extremely important outlet for rubbers. During the last 20 years still further major improvements have been made to the original styrene–butadiene, and new synthetic rubbers, based on polybutadiene and polyisoprene, which more nearly match natural rubber in both chemical constitution and physical properties, have reached full scale commercial production. In the same period continuous progress has been made in the production of new types and in the improvement of existing special-purpose rubbers to meet the ever-growing needs of industry. Thus, although natural rubber is still an important raw material, all major industrial countries have synthetic rubber available and are no longer completely dependent on the natural product.

U.K. RUBBER CONSUMPTION AND PRODUCTION

The present position in the United Kingdom is that the International Synthetic Rubber Co. Ltd., a consortium formed by the major tyre companies, produces styrene–butadiene rubber (SBR) at Hythe, Nr. Southampton with a capacity of 130,000 t/a and polybutadiene rubber (BR) at Grangemouth, Scotland with a capacity of 50,000 t/a; polyisoprene rubber is not yet made in the United Kingdom. Neoprene is produced by DuPont in Northern Ireland and isobutene–isoprene ("Butyl") rubber by Esso at Fawley, Hants, both plants having capacities of about 30,000 t/a each. Butadiene–acrylonitrile rubber is made on a 7500 t/a scale by B.P. Chemicals (U.K.) Ltd. at Barry, Glam., and several companies, including Dunlop, I.C.I. and Uniroyal, produce relatively small quantities of special-purpose rubbers, mainly in the form of latices. United Kingdom consumption of natural and synthetic rubbers over the past 8 years is shown in Table 24.

TABLE 24. U.K. CONSUMPTION OF NATURAL AND
SYNTHETIC RUBBERS
(figures in 000 tons)

Year	Natural	Synthetic	Total
1960	180	116	296
1961	166	121	287
1962	164	133	297
1963	169	143	312
1964	181	165	346
1965	183	183	366
1966	181	193	374
1967	176	196	372

RAW MATERIALS

A number of important synthetic rubber monomers are also used in large quantities for plastics and have already been described; styrene is dealt with on p. 187, acrylonitrile on p. 196, ethylene and propylene on p. 150 and isobutene on p. 176. This leaves three monomers to be described here—butadiene, 2-chlorobutadiene and isoprene.

Butadiene

There are two isomeric butadienes with the double bonds in the 1,2- and the 1,3-positions. The 1,2-isomer is of no importance industrially and the word butadiene is always understood as referring to the conjugated compound, 1,3-butadiene. This has a boiling point of $-5°C$ and is, therefore, a gas at normal temperatures and pressures. It may be stored as a liquid at slightly elevated pressures.

In the special conditions prevailing in Germany immediately before and during the Second World War and in the rush to produce synthetic rubber in the United States after the fall of Malaya, a number of routes to butadiene were used which would not now be considered economic. These are briefly described in the series of reaction schemes below:

1. From ethanol via acetaldehyde, acetaldol and 1,3-butanediol.

$$2C_2H_5OH \longrightarrow 2CH_3CHO \longrightarrow CH_3CHOHCH_2CHO$$

$$CH_3CHOHCH_2CHO \xrightarrow{+H_2} CH_3CHOHCH_2CH_2OH$$

$$CH_3CHOHCH_2CH_2OH \xrightarrow{-2H_2O} CH_2{=}CHCH{=}CH_2$$

2. From formaldehyde and acetylene via 1,4-butanediol.

$$2HCHO + CH{\equiv}CH \longrightarrow HOCH_2C{\equiv}CCH_2OH$$

$$HOCH_2C \equiv CCH_2CH \xrightarrow{+ 2H_2} HOCH_2CH_2CH_2CH_2OH$$

$$HOCH_2CH_2CH_2CH_2OH \xrightarrow{-2H_2O} CH_2{=}CHCH{=}CH_2$$

3. By simultaneous dehydration and dehydrogenation of ethanol.

$$2C_2H_5OH \xrightarrow{\quad -2H_2O-H_2 \quad} CH_2=CHCH=CH_2$$

One or other of these processes may still be used in some parts of the world for production of relatively small quantities of butadiene but the great bulk of the hydrocarbon is now derived from the mixed C_4 hydrocarbons produced as a by-product of naphtha cracking for ethylene or is made by the catalytic dehydrogenation of butene or butane. Butane dehydrogenation is not practised in the United Kingdom but Esso have a butene dehydrogenation plant at Fawley, near Southampton, with a capacity of 40,000 t/a. Esso also recover the by-product butadiene from their ethylene crackers as also do I.C.I., B.P. and Shell. The total U.K. output from both types of operation is currently about 120,000 t/a.

2-Chloro-1,3-butadiene (chloroprene)

This monomer is a liquid boiling at 59°C. The manufacturing process in use for many years starts with dimerization of acetylene to monovinyl acetylene; this reaction takes place at room temperature, at 10 atm pressure, in the presence of mixed cuprous and ammonium chlorides as catalyst. The monovinyl acetylene is separated from any co-produced divinyl acetylene and is then reacted with aqueous hydrogen chloride in the presence of the same catalyst.

$$2CH\equiv CH \longrightarrow CH\equiv CCH=CH_2 \xrightarrow{\quad +HCl \quad}$$
$$CH_2=CClCH=CH_2$$

DuPont have a plant operating this process near Londonderry in Northern Ireland, using acetylene from an adjoining plant of British Oxygen Chemicals Ltd. which was specially built to supply DuPont's needs.

Another route, recently developed, starts from n-butenes and butadiene and employs a series of operations of chlorination followed by pyrolytic dehydrochlorination. From butadiene the reactions are:

$$CH_2=CHCH=CH_2 + Cl_2 \longrightarrow CH_2ClCH=CHCH_2Cl +$$
$$CH_2=CHCHClCH_2Cl$$
$$CH_2ClCH=CHCH_2Cl \xrightarrow{\text{isomerization}} CH_2=CHCHClCH_2Cl$$
$$CH_2=CHCHClCH_2Cl \xrightarrow{\text{- HCl}} CH_2=CHCCl=CH_2$$

This process is used in France but was developed by the D.C.L. (now B.P. Chemicals Ltd.).

Isoprene

Isoprene is a liquid boiling at 34°C; although laboratory routes for its production have been known for many years, development of an economic process for large scale manufacture had to await the stimulation of a market demand for the hydrocarbon. Three processes with industrial potential are dimerization of propylene followed by pyrolysis in which methane is split off; a reaction between isobutene and formaldehyde followed by catalytic decomposition of the intermediate product; and partial hydrogenation of the reaction product between acetone and acetylene followed by dehydration. It seems probable, however, that most isoprene of the future will be produced by dehydrogenation of tertiary amylenes (pentenes) recovered from the C_5 fractions from refinery cracking operations. This source can be supplemented by dehydrogenation of isopentane, which is also available from refinery processes, if required. Isoprene is not yet produced on a large scale in the United Kingdom.

STYRENE–BUTADIENE RUBBERS (SBR)

The first synthetic rubber of this type was made by emulsion polymerization of a mixture of 75 parts of butadiene and 25 parts of styrene, at 50°C, using potassium persulphate as a catalyst. The product was well below the standard of the best natural rubber although it found widespread use when natural rubber was scarce, even in tyres.

Better techniques of polymerization were worked out in the early 1950's to give a greatly improved product. The main changes were the use of temperatures of about 5°C instead of 50°C for polymerization and the addition of a redox catalyst (see p. 25). This has the effect of favouring 1,4-addition of butadiene into the growing chain whereas, at the higher temperature, 1,2-addition predominates.

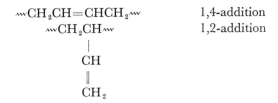

$\text{\textasciitilde\textasciitilde}CH_2CH{=}CHCH_2\text{\textasciitilde\textasciitilde}$ 1,4-addition
$\text{\textasciitilde\textasciitilde}CH_2CH\text{\textasciitilde\textasciitilde}$ 1,2-addition
         |
        CH
        ‖
        CH_2

The presence of the side chains in 1,2-addition and, in particular, the fact that the residual double bonds which are essential for the vulcanization process are present in the side chains, has an adverse effect on the properties of the resulting rubber.

An emulsion of butadiene and styrene is prepared in water containing the peroxide catalyst, cooled to 5°C and the Redox catalyst is added to this as it enters the reaction vessels, which are vigorously stirred. The process is a continuous one and the reactants pass through a chain of reactors until about 60% of the monomers have been converted to a latex; residence time in the system is 10–13 hr. A "short stop", usually a dithiocarbamate, is then added, excess butadiene removed by raising the temperature of the latex to 50°C and reducing the pressure and the unreacted styrene stripped off by steam distillation under vacuum. An antioxidant is added to the latex which is then coagulated with sodium chloride and sulphuric acid; the "crumb" is recovered by filtration, washed, dried and baled.

If the reaction is allowed to go beyond the 60% conversion stage it becomes uneconomically slow, owing to the decreased concentration of the reactants, and an excessive number of side chains is formed which, as noted above, is undesirable.

Elastomers with a range of properties can be made by suitably varying the conditions. It might, indeed, be said that SBR can be tailor-made to suit the end application.

Oil-Extended Rubbers

In the production of styrene–butadiene rubber some saturated polymer is found which acts as a diluent; the proportion decreases as the molecular weight increases. High molecular weight rubber is not easy to process but this can be overcome by extending it with oil, i.e. by the addition of an emulsion of heavy oil to the latex before coagulation. In this way an expensive diluent is, in effect, being replaced by a cheap one.

For many purposes the straight rubber is preferred, but oil-extended rubbers are now being used for a great variety of applications, including tyres. The proportion of oil can vary from 25 to 75 parts per 100 of rubber hydrocarbon.

Compounding

As with natural rubber and many plastics, styrene–butadiene rubber does not develop its optimum properties unless properly compounded; the processing techniques for it are very similar to those for natural rubber. These involve first masticating the rubber on a two-roll mill or in a closed mixer, such as a Banbury, followed by addition of a variety of reinforcing agents, antioxidants, accelerators, softeners, vulcanizing agents, dyestuffs, pigments and fillers according to the end use for which the rubber compound is required.

The most widely used reinforcing agent is carbon black, which may be added to an extent approaching at least half the amount of rubber present; even higher quantities are used for some applications; some other materials such as china clay and silica in various forms also have a reinforcing effect. The reinforcing action of these materials in improving the strength and abrasion resistance of the compound is most marked and, without them, the styrene–butadiene rubber would be virtually useless; they should not be confused with other inorganic additions such as whiting and barytes, which may be added to improve processing properties, hardness and chemical resistance or merely to cheapen the final product.

The accelerators speed up the process of vulcanization by sulphur which would otherwise be too slow for large scale production methods. A variety of chemicals, mainly sulphur and nitrogen compounds, is used, of which mercaptobenzothiazole is typical. Accelerators are activated by zinc oxide and a fatty acid. Many of the applications of rubber require that it shall adhere well to fabrics and similar materials; this is especially true in tyre building. This property of adhesiveness is known in the industry as "tack" and is conferred or enhanced by the addition of softeners to the mix. These may be vegetable or mineral oils, bitumen and resins. A typical compound might be made up as shown in Table 25.

TABLE 25. COMPOSITION OF A TYPICAL
STYRENE–BUTADIENE RUBBER COMPOUND

	Parts by weight
Styrene–butadiene rubber	100
Carbon black	45
Zinc oxide	3
Sulphur	2
Stearic acid	0·5
Accelerator	1·2

Such a compound would be cured at $135°C$ under 45 lb/in^2 steam pressure for 40 min. It would then have a tensile strength of 3500 lb/in^2, elongation at break of 770% and modulus at 300% extension of 650 lb/in^2.

Properties and Applications

Styrene–butadiene rubber is broadly similar to the natural product, being slightly superior in some respects and inferior in others. Its applications, therefore, follow closely those of natural rubber with a slight preference for those uses where its superior properties can be exploited.

The largest tonnage outlet is in passenger car tyres since tyre treads containing a high proportion of SBR show superior abrasion resistance to those made from natural rubber. Continued flexing, however, produces a greater heat build-up than with natural rubber and this severely limits its use in heavy duty truck tyres.

SBR is not as good as natural rubber in tear strength but it has superior ageing properties. It has captured a substantial part of the market for rubber footwear and general moulded products and has now taken its place as a recognized commodity in the rubber industry. It is unlikely to make any major new penetration into established applications unless economic conditions should change to favour its use in outlets in which it is marginally inferior to other rubbers. The market for it can, however, be expected to grow with the present applications.

BUTYL RUBBERS

Isobutene can be polymerized to produce rubbery materials but, as noted on p. 177, these have no residual unsaturation and cannot be vulcanized. The most important material commercially is a copolymer of isobutene and isoprene, which is made by solution polymerization. A mixture of isobutene with 4–6% of isoprene is dissolved in an inert solvent, such as methyl chloride, and the solution cooled to $-90°C$; aluminium trichloride dissolved in methyl chloride is added as catalyst. There is no latex stage as the polymer is formed immediately in distinct crumb-like particles. The process is continuous and the suspension of rubber particles coming from the reactor is mixed with hot water which drives off the methyl chloride and any unreacted monomers; the crumb is drained, dried and pelletized before being extruded as a block. The only U.K. producer is Esso, who have a 30,000 t/a plant at Fawley, near Southampton.

Butyl rubber is compounded and vulcanized by similar techniques to those used for natural rubber but higher temperatures may be used with advantage. Carbon black is the most useful reinforcing agent, increasing both abrasion resistance and toughness although it has no effect on the tensile strength. Zinc oxide is usually incorporated but, unlike natural and some other synthetic rubbers, a fatty acid is not required. Vulcanization may be carried out using sulphur in the usual way but, owing to the very low residual unsaturation, very active accelerators such as thiuram sulphides or dithiocarbamates and a high vulcanizing temperature around 180°C are required to give a reasonable curing time. Sulphur is not the only vulcanizing material for butyl rubber, and other methods, such as the use of nitrogen derivatives of quinone in the presence of an oxidizing agent, have been employed. A disadvantage of butyl rubber is that, unlike SBR, it is not compatible with natural rubber.

Properties and Uses

Compared with styrene–butadiene rubber, butyl rubber has a higher tensile strength around 4500 lb/in². It has about the same resistance to solvents, although it is more resistant to vegetable oils, but becomes brittle at low temperatures and softens on heating in an oxidizing atmosphere.

The unique property of butyl rubber is its extreme impermeability to gases and this, coupled with its resistance to ageing and tearing, has made it the material *par excellence* for tyre inner tubes. The gradual change over to tubeless tyres has brought about some reduction in demand for this purpose. It is generally recognized as one of the best materials for high voltage cable insulation, especially ships cables; it is also used for tank linings, chemical plant hose, moulded components and in the production of rubber cements and coated fabrics.

BUTADIENE RUBBER

1,4-Polymers of butadiene may have either a *cis-* or a *trans-* configuration as described for isoprene on p. 263. The development of stereospecific catalysts made it possible to manufacture polybutadienes of high *cis* content and the ready availability and low cost of butadiene made these appear superficially attractive; straight polybutadiene is, however, difficult to process. The polymers may be made in plant identical to that used for polyisoprene and the outline process diagram given on p. 276 would do equally well for polybutadiene.

The *cis* content of the polymer depends on the conditions of polymerization and on the catalyst used; a drop in *cis* content is associated with higher cold flow and more difficult processing although these properties are markedly affected also by the molecular weight and molecular weight distribution. Material of the highest *cis* content (about 96%) is made when aluminium alkyl chlorides are used in conjunction with cobalt salts while aluminium alkyls with titanium iodide produce a rubber of slightly lower *cis* content but with negligible reduction in properties. A product of *cis* content around 35% is made by using lithium butyl as catalyst.

Polybutadiene is vulcanized by sulphur, slightly less being required than for natural rubber, but it needs rather more accelerator. Reinforcing fillers are just as effective as with other rubbers, carbon black being the one most commonly used.

Properties and Applications

The tensile strength and tear resistance of butadiene rubber is lower than for natural rubber but it has rather better resistance to oxidation and ageing. It is always used blended with other rubbers and has found a substantial outlet in tyres, blended with natural rubber for heavy duty tyres and with SBR for car tyres. The abrasion resistance of these blends is excellent.

Other outlets for butadiene rubber are small. It is used to some extent in high impact polystyrene, as an alternative to SBR, and in belting where its high resistance to abrasion is of value.

ISOPRENE RUBBER

The *cis-* and *trans*-configuration of the 1,4-polymers of isoprene has been described on p. 263. As with polybutadiene the production of synthetic polymers of high *cis* or *trans* content followed the discovery of the Ziegler–Natta stereospecific catalysts. Polyisoprenes cannot be made by emulsion polymerization since the catalysts are destroyed by water and it is necessary to use a solution process, water being rigorously excluded from the system. Butyl lithium or the Ziegler aluminium alkyl–titanium tetrachloride catalysts are mainly used in solution in a hydrocarbon; the aluminium alkyl may be replaced by aluminium hydride.

The polymerization reaction may take from 3 to 6 hr at temperatures ranging between 5° and 40°C. The rubber solution so obtained is blended with a polymerization stopper and an antioxidant, washed with water to remove the catalyst and the solvent and any unreacted isoprene distilled off. The rubber crumbs are separated from any residual water on a vibrating screen and dried. An outline production scheme is shown in Fig. 23. By changing catalysts and process details, polymers of varying molecular weight and molecular weight distribution can be obtained.

Commercial isoprene rubbers are not identical with the natural product since the attainment of virtually 100% *cis* content, although theoretically possible, would not be justified on grounds of cost. Synthetic polymers of *cis* content from 92 to 96%, which differ little from natural rubber, can be made and sold at competitive and stable prices and have the advantage of more consistent quality. Synthetic isoprene rubber is steadily taking the place of natural rubber in a wide range of applications and is now making a significant penetration into the tyre market.

By modifications to process conditions and catalyst, polymers of high *trans* content, similar to natural gutta percha, may be made, and Dunlop do this in the United Kingdom for the manufacture of covers for golf balls.

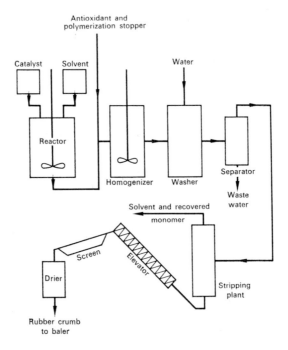

Fig. 23. Process scheme for isoprene rubber.

ETHYLENE–PROPYLENE RUBBERS

Linear polyethylene and polypropylene, as mentioned in earlier chapters, are partially crystalline and have a relatively high softening point. Blends of ethylene and propylene, for example in equimolecular proportions, polymerize to form amorphous and elastomeric products. They differ from the elastomers previously described, however, in being completely saturated.

Polymerization takes place in a hydrocarbon solvent at temperatures between 30° and 60°C and at pressures of 50–150 lb/in². A typical catalyst is aluminium triethyl and vanadium tetrachloride.

These elastomers have the advantage of better resistance to oxidation than those with residual unsaturation in the molecule but, because of their saturated character, they cannot be vulcanized by sulphur. Crosslinking can, however, be brought about by organic peroxides at high temperatures and in the absence of air. Complications arise with this method of curing, which is relatively expensive, when the product is highly filled and/or extended with oil.

The chief uses of ethylene–propylene rubbers are for cable insulation, extruded weatherstrip for buildings, cars, automotive mouldings, sheeting and conveyor belting.

In theory these should be very cheap rubbers but the nonstandard method of vulcanization and the various disadvantages arising from this have resulted in only slow development.

The curing problem can be overcome by the introduction of a double bond into the rubber molecule (cf. butyl rubber) by copolymerizing with a diene such as cyclopentadiene. The diene must be so chosen that the double bond is in a branch chain so that attack on it does not result in rupture of the main chain. Such terpolymers tend to be expensive since the diene is more costly than the simple olefins; the polymerization is a sophisticated operation but precise details of the methods used are not available. Oxidation resistance is slightly impaired compared with straight ethylene–propylene rubbers but the terpolymers can be processed with fillers and oils without adversely affecting the vulcanization efficiency.

As a family of elastomers the ethylene–propylene copolymers are very interesting and, although they are now coming into use in significant quantities commercially (especially in the United States), it is still difficult to assess their future. There is no sound evidence yet to suggest that they will achieve large scale use for tyres.

Ethylene–propylene rubbers are not yet made in the United Kingdom but I.S.R. are bringing a plant on stream at Grangemouth to make up to 10,000 t/a of these materials. No details of the process to be used or of the type of polymer to be made are available at the time of writing (August 1968).

BUTADIENE–ACRYLONITRILE RUBBER

The synthetic rubbers so far described have all been essentially replacements for natural rubber in its normal uses. With the butadiene–acrylonitrile copolymers we enter the field of special rubbers which can be used in applications where natural rubber is unsuitable and especially where resistance to hydrocarbon oils is necessary.

The rubber was discovered in Germany in 1930 and commercialized in 1937 as Buna N. It is normally produced by copolymerizing a mixture of the monomers, containing about 25% of acrylonitrile, by an emulsion process at temperatures between 5° and 35°C and pressures of 25–30 lb/in^2; conditions may be varied according to the grade of rubber required. The polymerization is slow and it may take up to 18 hr to reach the required stage. The latex formed is "dropped" from the reaction vessel at a carefully controlled point, lauryl mercaptan as a chain stopper and an antioxidant such as phenyl betanaphthylamine are added, and any residual monomers recovered by heating; these are recirculated.

Much of the product is required in the form of latex but, where solid rubber is required, the latex is heated, acid brine is added and the precipitated crumbs are washed with alkali and sheeted. The polymer is slightly cross-linked and so has some of the properties of a scorched rubber; this means that more softeners are necessary to give the required processing characteristics. Vulcanization is conventional and carbon black acts as a reinforcing agent.

Butadiene–acrylonitrile rubbers are equal to natural rubber in tensile strength but inferior in adhesion and tear resistance. Their outstanding advantage over the natural product is their excellent resistance to oil, solvents and many chemicals; the higher the acrylonitrile content the greater the oil resistance, but the more leathery the product begins to feel.

The latices are used in formulation of adhesives (see p. 82), as finishing agents for textiles and for impregnating both paper and textiles. The solid rubber is used for the manufacture of carburettor and fuel pump diaphragms in motor cars, for aircraft hoses and gaskets, for the hoses for petrol station pumps and for insulating oil-resistant cables. It also finds some use in the production of flexible friction materials.

Butadiene–acrylonitrile rubbers are often used in blends with other resins; thus, with novolaks (see p. 82), their processing properties and tack at elevated temperatures are improved, while the cured blends show much better abrasion resistance and some increase in tensile strength and tear resistance.

These rubbers are completely compatible with PVC and the blends can be processed on conventional rubber or plastics machinery. The rubber acts as a non-migratory plasticizer (see p. 210) in the PVC while the PVC increases the tensile strength and resistance to tearing of the rubber. The proportions may vary but about 50 parts of PVC to 100 parts of rubber are usual.

CHLOROPRENE RUBBER

Polymers of 2-chlorobutadiene (chloroprene) are often referred to by the original DuPont trade name of Neoprene. They behave rather differently from other synthetic rubbers during manufacture and processing. The monomer polymerizes slowly in the dark to produce an elastic mass, resembling vulcanized rubber, which is not thermoplastic and cannot be worked on a mill or sheeted; in commercial manufacture the polymerization is stopped at an intermediate stage to give a product which can be processed and cured.

Normally the monomer is emulsion polymerized under slightly alkaline conditions at about 40°C, using potassium persulphate as an initiator and with about 0·6% of sulphur added to the mix to prevent the reaction going too far. When the polymerization has reached the desired stage an emulsion of tetraethyl thiuram disulphide is added; this breaks some of the sulphur linkages formed during the reaction and increases the plasticity of the separated polymer.

Some of the product is required in latex form but, when solid rubber is to be the end product, an unusual method of breaking the emulsion is employed because salt-mineral acid coagulation gives an intractable sticky mass rather than a crumb. Acetic acid is added at a point just short of coagulation and the process is completed by refrigerating the emulsion while a drum rotates in it. A film of rubber forms on the drum and is stripped off, washed, passed through squeeze rolls and dried in air at 120°C.

Structure and Curing

Chloroprene polymerizes by both 1,4- and 1,2-addition, the 1,2-linkages providing active chlorine atoms which take part in the curing process. Unlike other rubbers, polychloroprene can be cured by heat alone but the process is normally accelerated by addition of zinc and magnesium oxides. The reaction may be represented as follows:

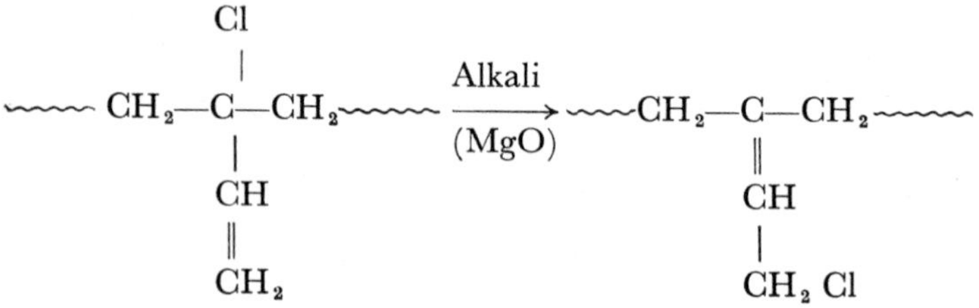

$$\sim\sim CH_2{-}\underset{\underset{\overset{\|}{CH}}{\underset{CH_2}{|}}}{\overset{\overset{Cl}{|}}{C}}{-}CH_2\sim\sim \quad\xrightarrow[\text{(MgO)}]{\text{Alkali}}\quad \sim\sim CH_2{-}\underset{\underset{\overset{|}{CH}}{CH_2\,Cl}}{\overset{\|}{C}}{-}CH_2\sim\sim$$

Two chains are then cross-linked by reaction with zinc oxide:

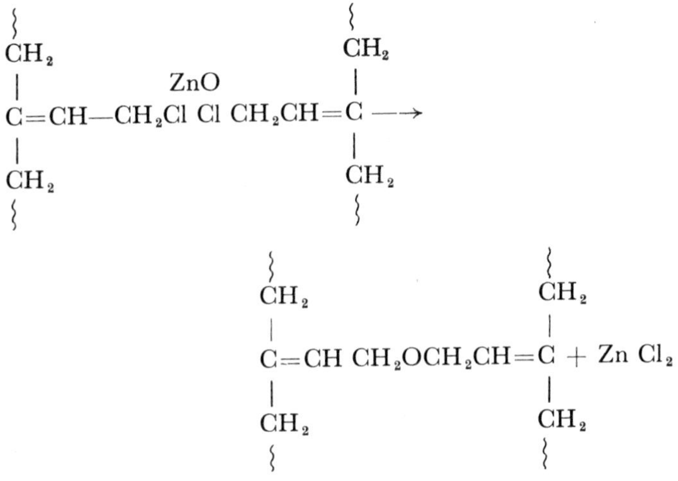

$$
\begin{array}{c}
\wr \\
CH_2 \\
| \\
C=CH-CH_2Cl \;\; Cl \; CH_2CH=C \;\longrightarrow \\
| \\
CH_2 \\
\wr
\end{array}
\qquad
\begin{array}{c}
\wr \\
CH_2 \\
| \\
\\
\\
| \\
CH_2 \\
\wr
\end{array}
$$

$$
\begin{array}{cc}
\wr & \wr \\
CH_2 & CH_2 \\
| & | \\
C=CH \; CH_2OCH_2CH=C & + \; Zn\;Cl_2 \\
| & | \\
CH_2 & CH_2 \\
\wr & \wr
\end{array}
$$

Properties and Applications

Chloroprene rubbers, like the nitrile rubbers (p. 278), have excellent resistance to oils and solvents; they combine this property with stability to the oxidizing effect of ozone, good ageing properties in heat and sunlight, excellent resistance to abrasion and a toughness and resilience comparable with natural rubber; the presence of chlorine in the molecule means that they are relatively non-flammable.

Uses for these rubbers are many and varied, although mostly for small quantities. They are widely used for engine mountings, gaskets, sealing rings and similar mechanical applications where long life and resistance to oil and grease are required. They may be employed for impregnation of various textiles for manufacture of protective clothing, conveyor belting and collapsible containers for the transport of oils and solvents; they are also used for cable insulation. Another substantial tonnage outlet is in the formulation of adhesives.

THERMOPLASTIC RUBBERS

It was noted in the introduction to this chapter that rubbers are a kind of plastic, intermediate between the thermosets and the true thermoplastics. They are, however, sufficiently like the thermosets to make it impracticable to remelt and remould them. Many thermoplastics show some elastomeric properties but, in general, their resilience is poor; plasticized PVC is probably the most rubber-like of the common thermoplastics and has been used as a substitute for rubber to an appreciable extent in certain applications, such as floor covering, wire coating, hose and belting.

It has recently been found possible to produce block copolymers of, for example, butadiene and styrene, which have the characteristics of rubbers but in which the polymer chains are sufficiently held together by van der Waals forces to give a "physical cross-link" at normal temperatures so that they need no vulcanization and are truly thermoplastic.

The materials can be produced by solution polymerization with an anionic initiator; one method is to use sodium methyl-styrene as the initiator and to carry out the reaction in tetra-hydrofuran solution at $-80°C$. The styrene is polymerized first and the butadiene or isoprene then added. The polymerization is stopped by addition of an acid, such as glacial acetic acid, and the rubber precipitated by addition of methanol, to which an oxidation inhibiter can conveniently be added.

Properties

These rubbers are normally available ready for use. They can be injection moulded or extruded on thermoplastics processing equipment, normally in the temperature range $140–220°C$ according to grade; the scrap can be recycled.

Thermoplastic rubbers can have tensile strengths similar to those of carbon black reinforced styrene–butadiene rubber. They retain their elastomeric properties between $-40°C$ and $+60°C$, but cannot be used at higher temperatures. In general most of their other properties are similar to those of other general purpose rubbers but, being unvulcanized, they are less resistant to organic solvents.

Their main outlets at present are in toys and sports goods and in miscellaneous mouldings and extrusions. Future developments seem likely to take these products into the adhesive and footwear fields.

OTHER SPECIAL-PURPOSE RUBBERS

The materials described make up the great bulk of commercially produced special-purpose rubbers. Other elastomeric polymers have been patented from time to time; the polysulphide rubbers were among the earliest materials to be produced on an industrial scale and are still produced in modest quantities today. They are made by reaction between an organic halogen compound, such as methylene chloride or ethylene dichloride, and sodium polysulphide and are made and marketed under the trade name "Thiokol" by the Thiokol Corporation in the United States. They are used as mastics and for sealing compounds.

Two other groups of elastomers, the polyurethanes and the silicones, are described in Chapters 7 and 11 respectively since it seemed better to group them with other polymers of the same chemical type. They have interesting properties but are relatively expensive and are used only in small quantities.

FUTURE DEVELOPMENT

There is now a considerable range of synthetic general-purpose rubbers which can compete with natural rubber from the economic standpoint and, in some cases, displace it on grounds of performance. Development seems likely to proceed on the lines of some economies in the production and processing of the currently expensive elastomers and, more particularly, in the production of new materials, or new grades of existing materials, especially suited for specific applications. The emergence of a new general-purpose material seems unlikely, although the comparatively recent discovery of the ethylene–propylene copolymers suggests that the possibility cannot be entirely ruled out.

In the special-purpose rubber field most efforts are likely to be devoted to development of materials which will retain their properties over a wider temperature range than those at present available. The outlets for such products, though small in quantity, will normally bear the high price which they are likely to command. Many of these outlets are likely to be found in the field of space vehicles and supersonic aircraft and in the advanced engineering techniques required in nuclear power generation and similar fields.

READING LIST

Introduction to Natural and Synthetic Rubber, by D. W. Huke, Hutchinson, London, 1961.

British Rubber Manufacturing, by Donnithorne, Duckworth, London.

Modern Synthetic Rubber, by Barron, Chapman and Hall, London, 1949.

Synthetic Rubber, by Whitby, John Wiley, New York, 1954.

Elastomers and Stereospecific Polymerization, American Chemical Society Symposium, 1966.

Chemistry of Natural and Synthetic Rubbers, by H. L. Fische, Reinhold, New York, 1953.

Conclusion

IN THE preceding chapters the authors have tried to give as full an account as possible, within the space available, of the modern plastics industry up to the stage of manufacture immediately preceding the production of articles for the final consumer. From an overall survey of this information some conclusions can be drawn on the way in which the industry is likely to develop in the future.

In spite of the relatively short period of 20 years during which the major growth of the industry has taken place, many of the developments were founded on work carried out in the two decades before the Second World War. It is clear that the polymerization of all the simple organic monomers has been fairly extensively studied and it seems unlikely that any new materials rivalling the three giants of the industry, PVC, polyethylene and polystyrene, in volume and price will emerge in the foreseeable future. Even polyformaldehyde, which has a very simple structure, appears unlikely ever to rival these three in price. It is also clear that some apparently unsatisfactory polymers have been made commercially successful by relatively minor modifications and this process may well continue to extend the range of usefulness of currently known materials.

Although they are only touched on in this book, it is the applications which determine the growth rate of output of plastics materials. Purely industrial applications are important but much of the growth of the industry is due to its output of modern consumer goods, the refrigerators, washing machines, vacuum cleaners, motor cars, radio and television receivers and similar products, which provide the amenities of life in the highly developed countries. A survey of world population indicates that there is an immense potential market open to the industry if the rate of world economic growth can be made to approach that of technological development more closely.

In the early days of plastics many difficulties were caused by attempts to use them directly as substitutes for traditional materials such as metal and wood. When the problem was approached by redesigning the product, having regard both to its function and to the properties of the material from which it was to be made, many of these difficulties disappeared. There is still room for progress in this field and this, together with market pressures, may well stimulate a large increase in output over the next decade. Thus the prospects for major increases in tonnage output can be summed up as more of the same or similar materials of improved quality.

NEW POLYMERS

In addition to the kind of development work described above, much research into the production of new polymers is constantly going on. The main aims are to produce polymers of better temperature resistance, greater modulus of elasticity (stiffness) and better processing characteristics. Some effort is being devoted to obtaining materials of enhanced value in a special respect such as chemical resistance, electrical resistance, dimensional stability and similar properties. The possibility of making materials with uniform characteristics over a wide temperature range, especially at temperatures below 0°C, is also being investigated.

The amount of information available on new polymers is obviously limited. Any company with a promising new product is unlikely to publish anything about it until either commercial development is so far advanced that complete secrecy is no longer possible or they have satisfied themselves that it is not going to be a major commercial success. The information on new polymers that is published does, however, give an indication of the lines which research is taking and this section will conclude with a brief account of some of the more interesting developments.

Polyimides

The six-membered benzene ring is known to be one of the most stable structures in organic chemistry, and much of the search for polymers of improved thermal stability has been concentrated on high molecular weight aromatic ring compounds. The polyimides are one result of this work; they are produced by the polycondensation of an aromatic dianhydride with an aromatic amine. As an example, 1,2,4,5-tetramethylbenzene (durene) may be oxidized to form the dianhydride, pyromellitic anhydride; this will react with 4,4'-diaminodiphenyl ether to form a polyamic acid. On heating the polyamic acid further ring closure takes place to form the polyimide. In order to simplify the printing of the following reaction scheme the two benzene rings and connecting oxygen link of the diaminodiphenyl ether has been designated by R.

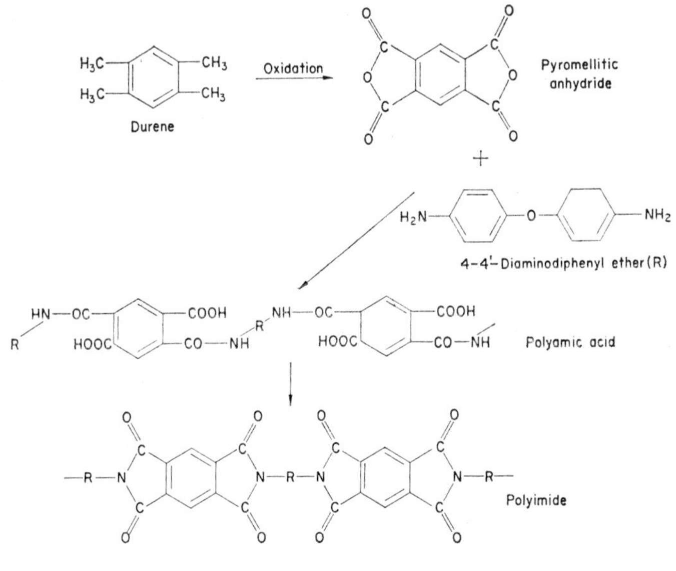

The softening point of the polymer is above 500°C, when some decomposition takes place, so that it cannot at present be moulded or extruded by conventional methods. Rods and various shapes can be produced by a sintering technique applied to polymer particles as is done with PTFE (see p. 223). The product is very rigid but impact strength is not very high; it is stable indefinitely at 260°C in air and at 315°C in a vacuum and may be heated intermittently to temperatures around 500°C without serious damage. The products can absorb water, as the sintering method of production gives a structure containing micro voids.

The polymer may be produced in film form by dissolving the polyamic acid in dimethyl formamide and casting a film from this solution. The polymerization is completed by heating this film with a monocarboxylic acid anhydride, such as benzoic or acetic anhydride, in the presence of a tertiary amine at 250–300°C. The high temperature resistance of the film makes it useful for printed circuits. Laboratory work on various combinations of anhydrides and amines has produced films with softening points as high as 800°C but these have not yet reached the stage of large scale manufacture.

A somewhat similar material to the one described above can be produced by pyrolysis of polyacrylonitrile when ring closure takes place, accompanied by dehydrogenation.

Polyacrylonitrile

Black orlon

This polymer is known as black "Orlon" and is a typical example of a so-called ladder polymer containing two chains, cross-linked at intervals. This structure ensures that, if one of the chains is broken, the molecular weight of the polymer does not immediately fall as it does in the case of the conventional straight chain polymers.

The material can be produced in the form of a thread which, when felted and laminated under pressure with a phenol–formaldehyde resin, forms a stiff board. When this board is graphited at 2000°C it forms a hard plate of virtually pure carbon which is believed to have some important uses in spacecraft.

Polyphenylene Oxide

A typical example of this type of polymer may be prepared by oxidation of 2,6-dimethyl phenol in the presence of a cupric chloride-pyridine complex as a catalyst; the copper salt must be present in excess, otherwise quinone formation takes place. The polymer chain is:

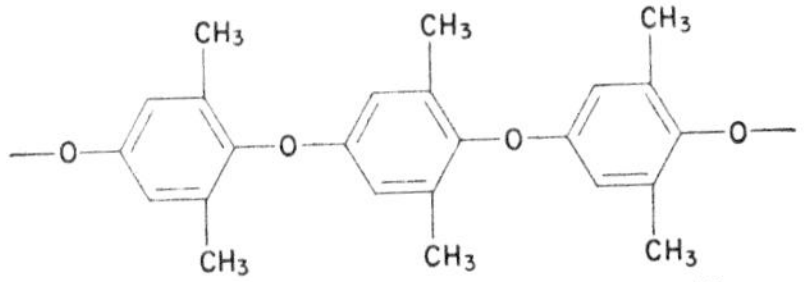

Polyphenylene oxide

A material of this type is produced by General Electric in the United States and by the A.K.U. Group in Holland. It has a high heat distortion temperature, up to 175°C, coupled with good dimensional stability. Injection mouldings can be made but high temperatures and high speeds of injection are required.

A modified product with a lower heat distortion temperature, around 130°C, has recently been introduced by A.K.U. Its other properties are similar to the original polyphenylene oxide. It is believed to be based on ortho-cresol instead of 2,6-dimethyl phenol and is considerably cheaper.

Polysulphones

Polysulphones are produced by condensing 4,4′-dichloro-diphenyl sulphone with the sodium salt of diphenylol propane. The resulting material is a straight chain polymer similar in structure to the polyphenylene oxide described in the previous section. Its formula may be written:

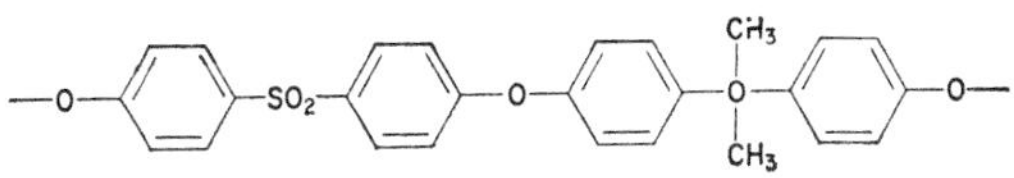

Polysulphone

The polysulphones have good mechanical properties, heat-distortion temperatures of over 150°C and can be processed at high temperatures in conventional equipment; they are non-flammable.

Union Carbide are currently marketing a polysulphone polymer and it is believed that I.C.I. are developing a somewhat similar product.

Polyparaxylene (Parylene)

When paraxylene is pyrolysed at 950°C in the presence of steam, the first reaction is believed to be dehydrogenation with the formation of *p*-xylilene; on quenching the reaction products this dimerizes to give di-*p*-xylene. If di-*p*-xylene is sublimed at low pressure it breaks up to give two *p*-xylilene radicals which immediately polymerize to form polyparaxylene. The reactions may be written:

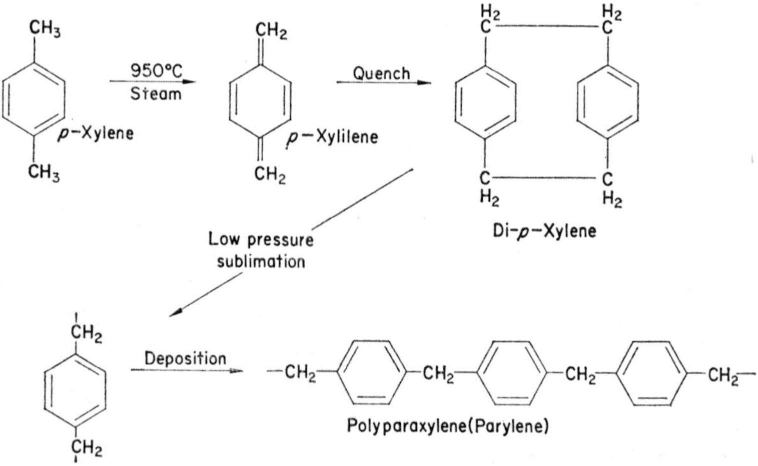

The polymer has outstanding electrical properties and much greater heat resistance than polystyrene; a film of the polymer has been shown to be stable for ten years in an inert atmosphere at 220°C. The development and commercial exploitation of this polymer is being undertaken mainly by Union Carbide in the United States.

Ionomers

These consist basically of a polyethylene chain with short branch chains attached. The branch chains terminate in a carboxyl group, the negative hydrogen having been replaced with a cation such as sodium, potassium, zinc or magnesium; with the bivalent metals some cross-linking is obtained. These products behave in some ways like thermosetting resins but the crosslinking becomes diffused at high temperatures and they can be processed by conventional thermoplastic techniques. They are remarkable in being completely transparent.

These materials have a low tensile strength and high elongation at break. The impact strength is very high but temperature resistance is poor; they show less stress cracking than polyethylene. DuPont have started marketing these polymers on a modest scale under the trade name "Surlyn".

Phosphonitrilic Polymers

This group of polymers has been studied over a long period but none of them has ever quite attained the status of a commercial product. They are rubbery materials and are stable at temperatures up to 350°C *in vacuo*. At the present stage of development their major disadvantage is instability in the presence of moisture, due to slow hydrolysis.

The polymers are made by reacting phosphorus pentachloride with ammonium chloride in the dry state or in tetrachloroethane as a solvent. A mixture of linear and cyclic polymers is first obtained which can be readily separated by dissolving the latter in an aliphatic hydrocarbon solvent. The cyclic polymer can be converted to the linear form by heating it with more phosphorus pentachloride.

The rubbery polymer is obtained by heating the linear material to temperatures of 200°C or above, depending on the molecular weight range of the starting material. It is believed to consist of long chains of $PNCl_2$ units cross-linked at intervals. This material is remarkable in being the only purely inorganic polymer even to have approached production on a commercial scale.

MISCELLANEOUS DEVELOPMENTS

Phosphorus pentachloride will condense with diphenylol propane to form a glass-like substance, melting between 60° and 120°C, which is believed to be the basis of a new resin which I.C.I. is developing under the name "Phoryl".

Many attempts have been made to produce useful polymers containing boron. These could be expected to be resistant to high temperature; the availability of boron hydrides, now being produced in substantial quantities for the atomic energy industry, has facilitated the work. Some of the most interesting compounds are the carboranes, a simple example being formed by interaction of decarborane and adiponitrile. The repeating unit in the polymer chain is:

$$-[-NC(CH_2)_4CNB_{10}H_{12}-]-$$

Some of these materials show considerable temperature resistance but, so far, none has achieved commercial scale production.

Polymers based on aluminium are also being widely studied, the starting point being the reaction between aluminium trihydride and various organic compounds, but they have not yet reached large scale production.

SUMMARY

The search for improved polymers appears to be concentrated on two areas of research. The first is a search for purely organic polymers and has been moderately successful; the most promising compounds are cyclic polymers containing many condensed aromatic nuclei. This line of research is not yet exhausted and further progress can be expected in the future. The second line of attack is the introduction of other elements, especially the tri- and penta-valent elements phosphorus, boron and aluminium, which are fairly plentiful in the earth's crust. Some progress has been made in the production of heat-resistant polymers but, as a class, these materials are rather unstable to moisture and tend to undergo slow hydrolysis.

There is at present no indication that a new star is likely to appear suddenly and future progress will be mainly by making better use of existing materials coupled with a slow but continuous improvement in their properties. Any new polymers developed, at least for some years to come, will probably find their main outlets in highly specialized fields which can bear their high cost.

READING LIST

Plastics Institute Transactions, June 1967, p. 509.
Chemical and Engineering News, 24th August 1964, p. 24.
Chemical Society Quarterly Reviews, 1964, vol. 18, p. 168.
Angewandte Chemie, November 1968, p. 835.
   Information on the properties of new materials, when commercial samples are available, is also given in *The Modern Plastics Encyclopaedia* each year but virtually no chemical information is supplied.

Name Index

A.K.U. 13
Albert, K. 8, 78
Albright & Wilson 236
American Cyanamid 10, 12, 86
Anchor Chemical Co. 195
A.S.T.M. 13, 49

Baekeland, L. 8, 39
Bakelite Ltd. 10, 60
Bakelite Xylonite Ltd. 10, 11, 12, 13, 60, 72, 78, 122, 152, 153, 155, 180, 195, 274
B.A.S.F. 10, 179
Bayer 134, 145, 228
Bexford Ltd. 12
(A.) Boake Roberts & Co. Ltd. 179
Board of Trade 2, 6
Borden Chemical Co. Ltd. 10, 122, 225
British Celanese Ltd. *see* Courtaulds Ltd.
British Cellophane Ltd. *see* Courtaulds Ltd.
British Cyanides Ltd. 11, 84
British Geon Ltd. 11, 201
British Industrial Plastics Ltd. 11, 13, 84, 96, 231
British Oxygen Co. Ltd. 11, 86, 267
B.P. Chemicals Ltd. *see* British Petroleum Co. Ltd.
B.P. Chemicals (U.K.) Ltd. *see* British Petroleum Co. Ltd.
British Petroleum Co. Ltd. 11, 13, 72, 153, 156, 195, 201, 202, 256, 265
British Plastics Federation 6, 7
British Visqueen Ltd. 12

British Xylonite *see* Bakelite Xylonite Ltd.
B.S.I. 13, 49
B.X.L. *see* Bakelite Xylonite Ltd.
B.X. Plastics Ltd. *see* Bakelite Xylonite Ltd.
Bushing Co. Ltd. 11, 78

Carothers 134
Celanese Corporation 230
C.I.B.A. (A.R.L.) Ltd. 96, 122
Commercial Plastics Ltd. 11, 78
Corning Glass Works 235
Courtaulds Ltd. 11, 213, 246, 250
Cyanamid of Great Britain Ltd. *see* American Cyanamid

Damard Lacquer Co. 60
De la Rue 12
Devoe & Raynolds 122
Distillers Co. Ltd. 10, 11, 13, 60, 179, 201
Dow Chemical Co. 11, 134, 179, 236
Dunlop 134, 265, 276
Du Pont 12, 134, 135, 137, 145, 175, 230, 265, 267, 291

Eastman Kodak 7
Ellis, Carleton 97
Erinoid 180, 256
Esso 153, 265, 273

295

Subject Index